Adler's Individual Psychology and Related Methods

ADLER'S INDIVIDUAL PSYCHOLOGY AND RELATED METHODS

ISBN: 978-1-291-85951-5

Copyright © 2014 Andreas Sofroniou.

All rights reserved. No part of this book may be reproduced, stored in a retrieval system or transmitted in any form or by any means without the prior written permission of the author and publisher, except by a reviewer who may quote brief passages in a review to be printed in a newspaper, magazine or journal.

At the specific preference of the author, this work to remain exactly as the author intended, verbatim, without editorial input.

Copyright © 2014 Andreas Sofroniou

ADLER'S INDIVIDUAL PSYCHOLOGY AND RELATED METHODS

ISBN: 978-1-291-85951-5

Andreas Sofroniou

Adler's Individual Psychology and Related Methods

CONTENTS:	PAGE:
1. PIONEER MEDICAL SPECIALIST	**5**
1.1 Career Path	5
1.2 Practice and Theory	5
1.3 Inferiority Complex	5
1.4 Neo-Freudianism	6
1.5 Goal and Development	6
2. ADLER'S PSYCHOLOGY	**7**
2.1 Inferiority	7
2.2 Initial Differences	7
2.3 Self-realisation	8
2.4 Personality Development	8
2.5 Child Guidance	9
3. INDIVIDUAL PSYCHOLOGY	**10**
3.1 Theories	10
3.2 Normality	10
3.3 Inferiority Complex	11
4. PARALLEL CONTRIBUTIONS	**12**
4.1 Sigmund Freud	12
4.2 Psychoanalysis	12
4.3 Hysteria	12
4.4 Abreaction and Catharsis	13
4.5 Subject Matter	13
4.6 Physician and Patient	14
4.8 Depth Psychology	14
4.9 Cathexis	15
4.10 Pleasure and Pain Principle	15
4.11 Mental Topography	16
4.12 Parapraxes	16
4.13 Theoretical Basis	16
4.14 Censorship	17
4.15 Sexual Instincts	17
4.16 Oedipus Complex	18
4.17 Transference	18
4.18 Sublimation	19
4.19 Psychoanalytic Movement	19
4.20 Freud's Sexuality and Development	20
4.21 Phallic Stage	21
4.22 Sexual Development	22
4.23 Cosmopolitan Support	24
5. FORERUNNERS AND CONTRIBUTORS	**26**
5.1 Breuer, Josef	26
5.2 Jung, Carl	26

5.3 Early Life and Career.	26
5.4 Jung's Character of Psychotherapy	27
5.5 RANK, OTTO	29
6. ANALYTICAL CONCEPTS	*31*
6.1 Psychoanalysis	31
6.2 Hysteria Disorder	32
6.3 Anxiety	33
6.4 Libido	35
6.5 Id	35
6.6 Ego	36
6.7 Superego	36
6.8 Oedipus	37
6.9 Psychoanalytic Dreams	38
6.10 Disagreement	39
6.11 Unconscious	40
6.12 Levels of Consciousness	40
6.13 Psychotherapies	41
6.14 Psychiatry	42
6.15 Mental Hygiene	43
6.16 Child Development	44
6.17 Child Psychology	45
7. PSYCHOLOGY AND BRANCHES	*47*
7.1 Scientific Discipline	47
7.2 History of psychology	48
7.3 Logic, Philosophy of Psychology	48
7.4 Aristotle and Works on Psychology	49
7.5 Sense and Sensible	49
7.6 Interest in Psychology	50
7.7 Comparative Psychology	51
7.8 Behaviourism	52
7.9 Applied Psychology	53
7.10 Experimental Psychology	55
7.11 Industrial Psychology	55
7.12 Social Psychology	56
7.13 Developmental Psychology	56
7.14 Educational Psychology	57
7.15 Psychopathology	58
7.16 Social Psychology	58
7.17 Clinical Psychology	59
8. ADLERIAN APPROACH	*61*
8.2 Compensation	61
8.3 Withdrawal	61
8.4 Therapy	62
8.5 Continuing Influences	62
9. ADLER'S PERSONAL LIFE	*64*

9.1 Earlier Personal life	64
9.2 Career	64
9.3 Adlerian School of Psychology	66
9.5 Basic Principles	68
9.6 Approach to Personality	69
9.7 Psychodynamics	69
9.8 Constructivism	70
9.9 Spiritual Holism	70
9.10 Typology	71
9.11 Importance of Memories	72
9.12 Birth Order	73
9.13 Addiction	74
9.14 Homosexuality	75
9.15 Parent Education	76
9.16 Spirituality, Ecology and Community	77
9.17 Death and Cremation	78
10. ADLER'S WORK IN THE 21ST CENTURY	***79***
10.1 Absorption into Modern Psychology	79
10.2 Private Logic	79
10.3 Professional Training	79
10.4 Adler University	80
10.5 Adlerian Psychology	80
10.6 Community Partnerships	80
10.7 Child Guidance Centre	81
10.8 Social Exclusion	81
10.9 Public Safety and Social Justice	81
10.10 Mental Health	82
INDEX	***83***
BIBLIOGRAPHY	***86***

Adler's Individual Psychology and Related Methods

1. PIONEER MEDICAL SPECIALIST

1.1 Career Path

Alfred Adler (1870-1937) was born in Vienna, and was considered to be a pioneer Austrian psychiatrist. Initially he practised as an ophthalmologist but later turned to mental health/mental illness and became a prominent member of the psychoanalytical society that formed around Sigmund Freud.

Adler disagreed with Freud's theories of 'infantile sexuality' and proposed that power, not sex, was the main key. This is when he introduced the idea of the inferiority complex.

In 1911 he seceded from the group of psychoanalysts and developed his own concepts on 'Individual Psychology'. His idea to develop his own branch of psychology, which investigates the individual considered as different from others.

1.2 Practice and Theory

The theory of psychology for the individual was firstly explained in Adler's published book 'Practice and Theory of Individual Psychology'. His main contributions to psychology include the 'inferiority complex', and his special treatment of neurosis as the 'exploitation of shock'.

Adler as a psychiatrist, at first a disciple of Sigmund Freud, he came to disagree with Freud's idea that mental illness was caused by sexual conflicts in infancy, arguing that society and culture were equally, if not more, significant factors.

1.3 Inferiority Complex

It was in 1907 when he introduced the concept of the inferiority complex, asserting that the key to understanding both personal and mass problems was the sense of inferiority and the individual's striving to compensate for this.

In 1911 he and his followers formed their own school to develop the ideas of individual psychology, and in 1921 he founded the first child guidance clinic, in Vienna.

1.4 Neo-Freudianism

Some modern psychotherapists consider Adler to be one of the first neo-Freudians, a term for influential psychoanalytic writers inspired by, but differing from, Freud. Loosely, it includes Jung, Melanie Klein (1882-1960), and Wilhelm Reich (1895-1957).

Frequently, however, it is restricted to those who rejected Freud's model of human development as the manifestation of instinctual drives and their often unsatisfactory suppression through fear of retribution.

In contrast, they emphasized that individuals develop through a process of integration into a social and cultural world which may be benign rather than hostile. They also denied that early infant-parent relationships exclusively caused the healthy or unhealthy development of personality, and they did not accept Freud's doctrine of the unconscious mind.

1.5 Goal and Development

Alfred Adler viewed people's behaviour as explicable in terms of the goals they adopt and their general attempt to develop their capacities and overcome a sense of their own imperfection and inferiority.

In a similar way, the American Harry Stack Sullivan (1892-1949) saw personality as a matter of how people relate to others throughout their lives, with adolescence a particularly important time in development. His compatriot Erik Erikson (1902-79) shared a very similar psychoanalytic orientation. He also discussed developmental 'tasks' and crises that occur throughout life.

Adler's Individual Psychology and Related Methods

2. ADLER'S PSYCHOLOGY

2.1 Inferiority

Adler, Alfred – the psychiatrist whose influential system of individual psychology introduced the term inferiority feeling, later widely and often inaccurately called inferiority complex. He developed a flexible, supportive psychotherapy to direct those emotionally disabled by inferiority feelings toward maturity, common sense, and social usefulness.

Throughout his life Adler maintained a strong awareness of social problems, and this served as a principal motivation in his work. From his earliest years as a physician (M.D., University of Vienna Medical School, 1895), he stressed consideration of the patient in relation to his total environment, and he began developing a humanistic, holistic approach to human problems.

2.2 Initial Differences

About 1900 Adler began to explore psychopathology within the context of general medicine and in 1902 became closely associated with Sigmund Freud. Gradually, however, differences between the two became irreconcilable, notably after the appearance of Adler's *Studie über Minderwertigkeit von Organen* (1907; *Study of Organ Inferiority and Its Psychical Compensation*), in which he suggested that persons try to compensate psychologically for a physical disability and its attendant feeling of inferiority. Unsatisfactory compensation results in neurosis.

Adler increasingly downplayed Freud's basic contention that sexual conflicts in early childhood cause mental illness and he further came to confine sexuality to a symbolic role in human strivings to overcome feelings of inadequacy. Outspokenly critical of Freud by 1911, Adler and a group of followers severed ties with Freud's circle and began developing what they called individual psychology, first outlined in *Über den nervösen Charakter* (1912; *The Neurotic Constitution*). The system was elaborated in later editions of this work and in other writings, such as *Menschenkenntnis* (1927; *Understanding Human Nature*).

2.3 Self-realisation

Individual psychology maintains that the overriding motivation in most people is a striving for what Adler somewhat misleadingly termed superiority--*i.e.,* self-realization, completeness, or perfection. This striving for superiority may be frustrated by feelings of inferiority, inadequacy, or incompleteness arising from physical defects, low social status, pampering or neglect during childhood, or other causes encountered in the natural course of life.

Individuals can compensate for their feelings of inferiority by developing their skills and abilities, or, less healthily, they may develop an inferiority complex, which comes to dominate their behaviour. Overcompensation for inferiority feelings can take the form of an egocentric striving for power and self-aggrandizing behaviour at others' expense.

2.4 Personality Development

Each person develops his personality and strives for perfection in his own particular way, in what Adler termed a style of life, or lifestyle. The individual's lifestyle forms in early childhood and is partly determined by what particular inferiority affected him most deeply during his formative years. The striving for superiority coexists with another innate urge: to cooperate and work with other people for the common good, a drive that Adler termed the social interest.

Mental health is characterized by reason, social interest, and self-transcendence; mental disorder by feelings of inferiority and self-centred concern for one's safety and superiority or power over others. The Adlerian psychotherapist directs the patient's attention to the unsuccessful, neurotic character of his attempts to cope with feelings of inferiority. Once the patient has become aware of these, the therapist builds up his self-esteem, helps him adopt more realistic goals, and encourages more useful behaviour and a stronger social interest.

2.5 Child Guidance

In 1921 Adler established the first child-guidance clinic in Vienna, soon thereafter opening and maintaining about 30 more there under his direction. Adler first went to the United States in 1926 and became visiting professor at Columbia University in 1927. He was appointed visiting professor of the Long Island College of Medicine in New York in 1932. In 1934 the government in Austria closed his clinics. Many of his later writings, such as *What Life Should Mean to You* (1931), were directed to the general reader. H.L. and R.R. Ansbacher edited *The Individual Psychology of Alfred Adler* (1956) and *Superiority and Social Interest* (1964).

3. INDIVIDUAL PSYCHOLOGY

3.1 Theories

It is widely acceptable that Individual Psychology is a body of theories of the Austrian psychiatrist Alfred Adler, who held that the main motives of human thought and behaviour are individual man's striving for superiority and power, partly in compensation for his feeling of inferiority.

Every individual, in this view, is unique, and his personality structure--including his unique goal and ways of striving for it--finds expression in his style of life, this life-style being the product of his own creativity. Nevertheless, the individual cannot be considered apart from society; all important problems, including problems of general human relations, occupation, and love, are social.

3.2 Normality

This theory of 'uniqueness' led to explanations of psychological normality and abnormality: although the normal person with a well-developed social interest will compensate by striving on the useful side of life (that is, by contributing to the common welfare and thus helping to overcome common feelings of inferiority), the neurotically disposed person is characterized by increased inferiority feelings, underdeveloped social interest, and an exaggerated, uncooperative goal of superiority, these symptoms manifesting themselves as anxiety and more or less open aggression.

Accordingly, he solves his problems in a self-centred, private fashion (rather than a task-centred, common-sense fashion), leading to failure. All forms of maladjustment share this constellation. Therapy consists in providing the patient with insight into his mistaken life-style through material furnished by him in the psychiatric interview.

3.3 Inferiority Complex

The inferiority complex definition relates to a feeling of inferiority that is wholly or partly unconscious. The term has been used by some psychiatrists and psychologists, particularly the followers of the early psychoanalyst Alfred Adler, who held that many outstanding achievements, some antisocial behaviour, and other continuing aspects of the personality could be traced to overcompensation for this feeling.

The use of "complex" later gained acceptance to denote the group of emotionally toned ideas, partially or even wholly repressed, organized around and related to such feelings of inferiority. The term inferiority complex has lost much of its significance through imprecise, popular misuse--for example, as an attempt at a facile explanation of any show of ambition by a person of small physical stature.

4. PARALLEL CONTRIBUTIONS

4.1 Sigmund Freud

Freud, Sigmund – the Austrian neurologist, was the founder of psychoanalysis.

Freud may justly be called the most influential intellectual legislator of his age. His creation of psychoanalysis was at once a theory of the human psyche, a therapy for the relief of its ills, and an optic for the interpretation of culture and society. Despite repeated criticisms, attempted refutations, and qualifications of Freud's work, its spell remained powerful well after his death and in fields far removed from psychology as it is narrowly defined.

If, as the American sociologist Philip Rieff once contended, "psychological man" replaced such earlier notions as political, religious, or economic man as the 20th century's dominant self-image, it is in no small measure due to the power of Freud's vision and the seeming inexhaustibility of the intellectual legacy he left behind.

4.2 Psychoanalysis

In the years 1880-2 a Viennese physician, Dr. Josef Breuer (1842-1925), discovered a new procedure by means of which he relieved a girl, who was suffering from severe hysteria, of her various symptoms. The idea occurred to him that the symptoms were connected with impressions which she had received during a period of excitement while she was nursing her sick father.

He therefore induced her, while she was in a state of hypnotic somnambulism, to search for these connections in her memory and to live through the "pathogenic" scenes once again without inhibiting the affects that arose in the process. He found that when she had done this the symptom in question disappeared for good.

4.3 Hysteria

This was at a date before the investigations of Charcot and Pierre Janet into the origin of hysterical symptoms, and Breuer's discovery was thus entirely uninfluenced by them. But he did not pursue the matter any further at the time, and it was not until some 10 years later that he took it up again in collaboration with Sigmund Freud.

In 1895 they published a book, *Studien über Hysterie,* in which Breuer's discoveries were described and an attempt was made to explain them by the theory of *Catharsis.* According to that hypothesis, hysterical symptoms originate through the energy of a mental process being withheld from conscious influence and being diverted into bodily innervation (*"Conversion"*).

4.4 Abreaction and Catharsis

A hysterical symptom would thus be a substitute for an omitted mental act and a reminiscence of the occasion which should have given rise to that act. On this view, recovery would be a result of the liberation of the affect that had gone astray and of its discharge along a normal path (*"Abreaction"*).

Cathartic treatment gave excellent therapeutic results, but it was found that they were not permanent and that they were dependent on the personal relation between the patient and the physician. Freud, who later proceeded with these investigations by himself, made an alteration in their technique, by replacing hypnosis by the method of free association.

He invented the term *"psychoanalysis,"* which in the course of time came to have two meanings:

(1) A particular method of treating nervous disorders and

(2) The science of unconscious mental processes, which has also been appropriately described as "depth-psychology."

4.5 Subject Matter

Psychoanalysis finds a constantly increasing amount of support as a therapeutic procedure, owing to the fact that it can do more for certain classes of patients than any other method of treatment.

The principal field of its application is in the milder neuroses-- hysteria, phobias and obsessional states, but in malformations of character and in sexual inhibitions or abnormalities it can also bring about marked improvements or even recoveries. Its influence upon dementia praecox and paranoia is doubtful; on the other hand, in favourable circumstances it can cope with depressive states, even if they are of a severe type.

4.6 Physician and Patient

In every instance the treatment makes heavy claims upon both the physician and the patient: the former requires a special training and must devote a long period of time to exploring the mind of each patient, while the latter must make considerable sacrifices, both material and mental. Nevertheless, all the trouble involved is as a rule rewarded by the results.

Psychoanalysis does not act as a convenient panacea ("cito, tute, jucunde") upon all psychological disorders. On the contrary, its application has been instrumental in making clear for the first time the difficulties and limitations in the treatment of such affections.

4.7 Therapeutic Results

The therapeutic results of psychoanalysis depend upon the replacement of unconscious mental acts by conscious ones and are operative in so far as that process has significance in relation to the disorder under treatment.

The replacement is effected by overcoming internal resistances in the patient's mind. The future will probably attribute far greater importance to psychoanalysis as the science of the unconscious than as a therapeutic procedure.

4.8 Depth Psychology

Psychoanalysis, in its character of depth-psychology, considers mental life from three points of view: the dynamic, the economic and the topographical.

From the first of these standpoints, the *dynamic* one, psychoanalysis derives all mental processes (apart from the reception of external stimuli) from the interplay of forces, which assist or inhibit one another, combine with one another, enter into compromises with one another, etc. All of these forces are originally in the nature of *instincts*; that is to say, they have an organic origin.

4.9 Cathexis

They are characterised by possessing an immense (somatic) persistence and reserve of power ("*repetition-compulsion*"); and they are represented mentally as images or ideas with an affective charge ("*cathexis*"). In psychoanalysis, no less than in other sciences, the theory of instincts is an obscure subject.

An empirical analysis leads to the formation of two groups of instincts: the so-called "ego-instincts," which are directed towards self-preservation and the "object-instincts," which are concerned with relations to an external object. The social instincts are not regarded as elementary or irreducible.

Theoretical speculation leads to the suspicion that there are two fundamental instincts which lie concealed behind the manifest ego-instincts and object-instincts: namely

(*a*) Eros, the instinct which strives for ever closer union, and

(*b*) The instinct of destruction, which leads toward the dissolution of what is living.

In psychoanalysis the manifestation of the force of Eros is given the name "*libido.*"

4.10 Pleasure and Pain Principle

From the *economic* standpoint psychoanalysis supposes that the mental representations of the instincts have a cathexis of definite quantities of energy, and that it is the purpose of the mental apparatus to hinder any damming-up of these energies and to keep as low as possible the total amount of the excitations to which it is subject.

The course of mental processes is automatically regulated by the "*pleasure-pain principle*"; and pain is thus is some way related to an increase of excitation and pleasure to a decrease. In the course of development the original pleasure principle undergoes a modification with reference to the external world, giving place to the "*reality-principle*," whereby the mental apparatus learns to postpone the pleasure of satisfaction and to tolerate temporarily feelings of pain.

4.11 Mental Topography

Topographically, psychoanalysis regards the mental apparatus as a composite instrument, and endeavours to determine at what points in it the various mental processes take place. According to the most recent psychoanalytic views, the mental apparatus is composed of an *"id,"* which is the reservoir of the instinctive impulses, of an *"ego,"* which is the most superficial portion of the id and one which is modified by the influence of the external world, and of a *"super-ego,"* which develops out of the id, dominates the ego and represents the inhibitions of instinct characteristic of man.

Further, the property of consciousness has a topographical reference; for processes in the id are entirely unconscious, while consciousness is the function of the ego's outermost layer, which is concerned with the perception of the external world.

At this point two observations may be in place. It must not be supposed that these very general ideas are presuppositions upon which the work of psychoanalysis depends. On the contrary, they are its latest conclusions and are in every respect open to revision.

4.12 Parapraxes

Psychoanalysis is founded securely upon the observation of the facts of mental life; and for that very reason its theoretical superstructure is still incomplete and subject to constant alteration. Secondly, there is no reason for astonishment that psychoanalysis, which was originally no more than an attempt at explaining pathological mental phenomena, should have developed into a psychology of normal mental life.

The justification for this arose with the discovery that the dreams and mistakes (*"parapraxes,"* such as slips of the tongue, etc.) of normal men have the same mechanism as neurotic symptoms.

4.13 Theoretical Basis

The first task of psychoanalysis was the elucidation of nervous disorders. The analytical theory of the neuroses is based upon three ground-pillars: the recognition of

    (1) *"Repression,"*

(2) The importance of the sexual instincts and

(3) "*Transference.*"

4.14 Censorship

There is a force in the mind which exercises the functions of a censorship, and which excludes from consciousness and from any influence upon action all tendencies which displease it. Such tendencies are described as "repressed." They remain unconscious; and if the physician attempts to bring them into the patient's consciousness he provokes a "*resistance.*"

These repressed instinctual impulses, however, are not always made powerless by this process. In many cases they succeed in making their influence felt by circuitous paths, and the indirect or substitutive gratification of repressed impulses is what constitutes neurotic symptoms.

4.15 Sexual Instincts

For cultural reasons the most intensive repression falls upon the sexual instincts; but it is precisely in connection with them that repression most easily miscarries, so that neurotic symptoms are found to be substitutive gratifications of repressed sexuality. The belief that in man sexual life begins only at puberty is incorrect.

On the contrary, signs of it can be detected from the beginning of extra-uterine existence; it reaches a first culminating point at or before the fifth year ("early period"), after which it is inhibited or interrupted ("latency period") until the age of puberty, which is the second climax of its development. This double onset of sexual development seems to be distinctive of the genus Homo.

All experiences during the first period of childhood are of the greatest importance to the individual, and in combination with his inherited sexual constitution, form the dispositions for the subsequent development of character or disease. It is a mistaken belief that sexuality coincides with "genitality." The sexual instincts pass through a complicated course of development, and it is only at the end of it that the "primacy of the genital zone" is attained.

Before this there are a number of "pre-genital organisations" of the libido--points at which it may become "fixated" and to which, in the event of subsequent repression, it will return ("*regression*"). The infantile fixations of the libido are what determine the form of neurosis which sets in later. Thus the neuroses are to be regarded as inhibitions in the development of the libido.

4.16 Oedipus Complex

There are no specific causes of nervous disorders; the question whether a conflict finds a healthy solution or leads to a neurotic inhibition of function depends upon quantitative considerations, that is, upon the relative strength of the forces concerned. The most important conflict with which a small child is faced is his relation to his parents, the "*Oedipus complex*"; it is in attempting to grapple with this problem that persons destined to suffer from a neurosis habitually fail.

The reactions against the instinctual demands of the Oedipus complex are the source of the most precious and socially important achievements of the human mind; and this probably holds true not only in the life of individuals but also in the history of the human species as a whole. The super-ego, the moral factor which dominates the ego, also has its origin in the process of overcoming the Oedipus complex.

4.17 Transference

By "*transference*" is meant a striking peculiarity of neurotics. They develop toward their physician emotional relations, both of an affectionate and hostile character, which are not based upon the actual situation but are derived from their relations toward their parents (the Oedipus complex).

Transference is a proof of the fact that adults have not overcome their former childish dependence; it coincides with the force which has been named "suggestion"; and it is only by learning to make use of it that the physician is enabled to induce the patient to overcome his internal resistances and do away with his repressions. Thus psychoanalytic treatment acts as a second education of the adult, as a corrective to his education as a child.

Adler's Individual Psychology and Related Methods

4.18 Sublimation

Within this narrow compass it has not been possible to mention many matters of the greatest interest, such as the "*sublimation*" of instincts, the part played by symbolism, the problem of "*ambivalence,*" etc.

Nor has there been space to allude to the applications of psychoanalysis, which originated, as we have seen, in the sphere of medicine, to other departments of knowledge (such as Anthropology, the Study of Religion, Literary History and Education) where its influence is constantly increasing.

It is enough to say that psychoanalysis, in its character of the psychology of the deepest, unconscious mental acts, promises to become the link between Psychiatry and all of these other fields of study.

4.19 Psychoanalytic Movement

The beginnings of psychoanalysis may be marked by two dates: 1895, which saw the publication of Breuer and Freud's *Studien über Hysterie,* and 1900, which saw that of Freud's *Traumdeutung.*

At first the new discoveries aroused no interest either in the medical profession or among the general public. In 1907 the Swiss psychiatrists, under the leadership of E. Bleuler and C.G. Jung, began to concern themselves in the subject; and in 1908 there took place at Salzburg a first meeting of adherents from a number of different countries. In 1909 Freud and Jung were invited to America by G. Stanley Hall to deliver a series of lectures on psychoanalysis at Clark University, Worcester, Mass.

From that time forward interest in Europe grew rapidly; it showed itself, however, in a forcible rejection of the new teachings, characterised by an emotional colouring which sometimes bordered upon the unscientific.

The reasons for this hostility are to be found, from the medical point of view, in the fact that psychoanalysis lays stress upon psychical factors, and from the philosophical point of view, in its assuming as an underlying postulate the concept of unconscious mental activity; but the strongest reason was undoubtedly the general disinclination of mankind to concede to the factor of sexuality such importance as is assigned to it by psychoanalysis. In spite of this widespread opposition, however, the movement in favour of psychoanalysis was not to be checked.

Its adherents formed themselves into an International Association, which passed successfully through the ordeal of the World War, and at the present time comprises local groups in Vienna, Berlin, Budapest, London, Switzerland, Holland, Moscow and Calcutta, as well as two in the United States. There are three journals representing the views of these societies: the *Internationale Zeitschrift für Psychoanalyse, Imago* (which is concerned with the application of psychoanalysis to non-medical fields of knowledge), and the *International Journal of Psycho-Analysis.*

During the years 1911-3 two former adherents, Alfred Adler, of Vienna, and C.G. Jung, of Zürich, seceded from the psychoanalytic movement and founded schools of thought of their own. In 1921 Dr. M. Eitingon founded in Berlin the first public psychoanalytic clinic and training-school, and this was soon followed by a second in Vienna. For the moment these are the only institutions on the continent of Europe which make psychoanalytic treatment accessible to the wage-earning classes.

4.20 Freud's Sexuality and Development

To spell out the formative development of the sexual drive, Freud focused on the progressive replacement of erotogenic zones in the body by others. An originally polymorphous sexuality first seeks gratification orally through sucking at the mother's breast, an object for which other surrogates can later be provided. Initially unable to distinguish between self and breast, the infant soon comes to appreciate its mother as the first external love object.

Later Freud would contend that even before that moment, the child can treat its own body as such an object, going beyond undifferentiated autoeroticism to a narcissistic love for the self as such. After the oral phase, during the second year, the child's erotic focus shifts to its anus, stimulated by the struggle over toilet training. During the anal phase the child's pleasure in defecation is confronted with the demands of self-control. The third phase, lasting from about the fourth to the sixth year, he called the phallic.

Because Freud relied on male sexuality as the norm of development, his analysis of this phase aroused considerable opposition, especially because he claimed its major concern is castration anxiety.

To grasp what Freud meant by this fear, it is necessary to understand one of his central contentions. As has been stated, the death of Freud's father was the trauma that permitted him to delve into his own psyche. Not only did Freud experience the expected grief, but he also expressed disappointment, resentment, and even hostility toward his father in the dreams he analyzed at the time. In the process of abandoning the seduction theory he recognized the source of the anger as his own psyche rather than anything objectively done by his father.

Turning, as he often did, to evidence from literary and mythical texts as anticipations of his psychological insights, Freud interpreted that source in terms of Sophocles' tragedy *Oedipus Rex*.

The universal applicability of its plot, he conjectured, lies in the desire of every male child to sleep with his mother and remove the obstacle to the realization of that wish, his father. What he later dubbed the Oedipus complex presents the child with a critical problem, for the unrealizable yearning at its root provokes an imagined response on the part of the father: the threat of castration.

4.21 Phallic Stage

The phallic stage can only be successfully surmounted if the Oedipus complex with its accompanying castration anxiety can be resolved. According to Freud, this resolution can occur if the boy finally suppresses his sexual desire for the mother, entering a period of so-called latency, and internalizes the reproachful prohibition of the father, making it his own with the construction of that part of the psyche Freud called the superego or the conscience.

The blatantly phallo-centric bias of this account, which was supplemented by a highly controversial assumption of penis envy in the already castrated female child, proved troublesome for subsequent psychoanalytic theory. Not surprisingly, later analysts of female sexuality have paid more attention to the girl's relations with the pre-Oedipal mother than to the vicissitudes of the Oedipus complex.

Anthropological challenges to the universality of the complex have also been damaging, although it has been possible to redescribe it in terms that lift it out of the specific familial dynamics of Freud's own day. If the creation of culture is understood as the institution of kinship structures based on exogamy, then the Oedipal drama reflects the deeper struggle between natural desire and cultural authority.

Freud, however, always maintained the intra-psychic importance of the Oedipus complex, whose successful resolution is the precondition for the transition through latency to the mature sexuality, he called the genital phase. Here the parent of the opposite sex is conclusively abandoned in favour of a more suitable love object able to reciprocate reproductively useful passion.

In the case of the girl, disappointment over the non-existence of a penis is transcended by the rejection of her mother in favour of a father figure instead. In both cases, sexual maturity means heterosexual, pro-creatively inclined, genitally focused behaviour.

4.22 Sexual Development

Sexual development, however, is prone to troubling maladjustments preventing this outcome if the various stages are unsuccessfully negotiated. Fixation of sexual aims or objects can occur at any particular moment, caused either by an actual trauma or the blockage of a powerful libidinal urge. If the fixation is allowed to express itself directly at a later age, the result is what was then generally called a perversion.

If, however, some part of the psyche prohibits such overt expression, then, Freud contended, the repressed and censored impulse produces neurotic symptoms, neuroses being conceptualized as the negative of perversions. Neurotics repeat the desired act in repressed form, without conscious memory of its origin or the ability to confront and work it through in the present.

In addition to the neurosis of hysteria, with its conversion of affective conflicts into bodily symptoms, Freud developed complicated etiological explanations for other typical neurotic behaviour, such as obsessive-compulsions, paranoia, and narcissism.

These he called psychoneuroses, because of their rootedness in childhood conflicts, as opposed to the actual neuroses such as hypochondria, neurasthenia, and anxiety neurosis, which are due to problems in the present (the last, for example, being caused by the physical suppression of sexual release).

Freud's elaboration of his therapeutic technique during these years focused on the implications of a specific element in the relationship between patient and analyst, an element whose power he first began to recognize in reflecting on Breuer's work with Anna O. Although later scholarship has cast doubt on its veracity, Freud's account of the episode was as follows.

An intense rapport between Breuer and his patient had taken an alarming turn when Anna divulged her strong sexual desire for him. Breuer, who recognized the stirrings of reciprocal feelings, broke off his treatment out of an understandable confusion about the ethical implications of acting on these impulses.

Freud came to see in this troubling interaction the effects of a more pervasive phenomenon, which he called transference (or in the case of the analyst's desire for the patient, counter-transference). Produced by the projection of feelings, transference, he reasoned, is the reenactment of childhood urges cathected (invested) on a new object. As such, it is the essential tool in the analytic cure, for by bringing to the surface repressed emotions and allowing them to be examined in a clinical setting, transference can permit their being worked through in the present. That is, affective remembrance can be the antidote to neurotic repetition.

It was largely to facilitate transference that Freud developed his celebrated technique of having the patient lie on a couch, not looking directly at the analyst, and free to fantasize with as little intrusion of the analyst's real personality as possible. Restrained and neutral, the analyst functions as a screen for the displacement of early emotions, both erotic and aggressive. Transference onto the analyst is itself a kind of neurosis, but one in the service of an ultimate working through of the conflicting feelings it expresses.

Only certain illnesses, however, are open to this treatment, for it demands the ability to redirect libidinal energy outward. The psychoses, Freud sadly concluded, are based on the redirection of libido back onto the patient's ego and cannot therefore be relieved by transference in the analytic situation. How successful psychoanalytic therapy has been in the treatment of psychoneuroses remains, however, a matter of considerable dispute.

4.23 Cosmopolitan Support

Although Freud's theories were offensive to many in the Vienna of his day, they began to attract a cosmopolitan group of supporters in the early 1900s. In 1902 the Psychological Wednesday Circle began to gather in Freud's waiting room with a number of future luminaries in the psychoanalytic movements in attendance.

Alfred Adler and Wilhelm Stekel were often joined by guests such as Sándor Ferenczi, Carl Gustav Jung, Otto Rank, Ernest Jones, Max Eitingon, and A.A. Brill. In 1908 the group was renamed the Vienna Psychoanalytic Society and held its first international congress in Salzburg. In the same year the first branch society was opened in Berlin. In 1909 Freud, along with Jung and Ferenczi, made a historic trip to Clark University in Worcester, Mass.

The lectures he gave there were soon published as *Über Psychoanalyse* (1910; *The Origin and Development of Psychoanalysis*), the first of several introductions he wrote for a general audience. Along with a series of vivid case studies--the most famous known colloquially as "Dora" (1905), "Little Hans" (1909), "The Rat Man" (1909), "The Psychotic Dr. Schreber" (1911), and "The Wolf Man" (1918)--they made his ideas known to a wider public.

As might be expected of a movement whose treatment emphasized the power of transference and the ubiquity of Oedipal conflict, its early history is a tale rife with dissension, betrayal, apostasy, and excommunication. The most widely noted schisms occurred with Adler in 1911, Stekel in 1912, and Jung in 1913; these were followed by later breaks with Ferenczi, Rank, and Wilhelm Reich in the 1920s. Despite efforts by loyal disciples like Ernest Jones to exculpate Freud from blame, subsequent research concerning his relations with former disciples like Viktor Tausk have clouded the picture considerably.

Critics of the hagiographic legend of Freud have, in fact, had a relatively easy time documenting the tension between Freud's aspirations to scientific objectivity and the extraordinarily fraught personal context in which his ideas were developed and disseminated. Even well after Freud's death, his archivists' insistence on limiting access to potentially embarrassing material in his papers has reinforced the impression that the psychoanalytic movement resembled more a sectarian church than a scientific community (at least as the latter is ideally understood).

5. FORERUNNERS AND CONTRIBUTORS

5.1 Breuer, Josef

Josef Breuer (1842–1925) the Austrian physician and physiologist was acknowledged by Sigmund Freud and others as the principal forerunner of psychoanalysis. Breuer found, in 1880, that he had relieved symptoms of hysteria in a patient, (called Anna O. in his case study), Bertha Pappenheim, after he had induced her to recall unpleasant past experiences under hypnosis. He concluded that neurotic symptoms result from unconscious processes and disappear when these processes become conscious.

Breuer described his methods and results to Freud and referred patients to him. With Freud he wrote *Studien über Hysterie* (1895), in which Breuer's treatment of hysteria was described. Later disagreement on basic theories of therapy terminated their collaboration.

Breuer's earlier work had dealt with the respiratory cycle, and in 1868 he described the Hering-Breuer reflex involved in the sensory control of inhalations and exhalations in normal breathing. In 1873 he discovered the sensory function of the semicircular canals in the inner ear and their relation to positional sense or balance. He practiced medicine and was physician to many members of the Viennese medical faculty.

5.2 Jung, Carl

Carl Gustav Jung (1875–1961), the Swiss psychologist and psychiatrist founded analytic psychology, in some aspects a response to Sigmund Freud's psychoanalysis.

Jung proposed and developed the concepts of the extroverted and introverted personality, archetypes, and the collective unconscious. His work has been influential in psychiatry and in the study of religion, literature, and related fields.

5.3 Early Life and Career.

Jung was the son of a philologist and pastor. His childhood was lonely, though enriched by a vivid imagination, and from an early age he observed the behaviour of his parents and teachers, which he tried to resolve. Especially concerned with his father's failing belief in religion, he tried to communicate to him his own experience of God.

Though the elder Jung was in many ways a kind and tolerant man, neither he nor his son succeeded in understanding each other. Jung seemed destined to become a minister, for there were a number of clergymen on both sides of his family. In his teens he discovered philosophy and read widely, and this, together with the disappointments of his boyhood, led him to forsake the strong family tradition and to study medicine and become a psychiatrist. He was a student at the universities of Basel (1895-1900) and Zürich (M.D., 1902).

He was fortunate in joining the staff of the Burghölzli Asylum of the University of Zürich at a time (1900) when it was under the direction of Eugen Bleuler, whose psychological interests had initiated what are now considered classical researches into mental illness. At Burghölzli, Jung began, with outstanding success, to apply association tests initiated by earlier researchers.

He studied, especially, patients' peculiar and illogical responses to stimulus words and found that they were caused by emotionally charged clusters of associations withheld from consciousness because of their disagreeable, immoral (to them), and frequently sexual content. He used the now famous term complex to describe such conditions.

5.4 Jung's Character of Psychotherapy

The rest of his life was given over to the development of his ideas, especially those on the relation between psychology and religion. In his view, obscure and often neglected texts of writers in the past shed unexpected light not only on Jung's own dreams and fantasies but also on those of his patients; he thought it necessary for the successful prosecution of their art that psychotherapists become familiar with writings of the old masters.

Besides the development of new psychotherapeutic methods that derived from his own experience and the theories developed from them, Jung gave fresh importance to the so-called Hermetic tradition. He conceived that the Christian religion was part of a historic process necessary for the development of consciousness, but he thought that the heretical movements, starting with Gnosticism and ending in alchemy, were manifestations of unconscious archetypal elements not adequately expressed in the varying forms of Christianity.

He was particularly impressed with his finding that alchemical-like symbols could be found frequently in modern dreams and fantasies, and he thought that alchemists had constructed a kind of textbook of the collective unconscious. He drove this home in four large volumes of his *Collected Works*.

His historical studies aided him in pioneering the psychotherapy of the middle-aged and elderly, especially those who felt their lives had lost meaning. He helped them to appreciate the place of their lives in the sequence of history. Most of these patients had lost their religious belief; Jung found that if they could discover their own myth as expressed in dream and imagination they would become more complete personalities. He called this process individuation.

In later years he became professor of psychology at the Federal Polytechnical University in Zürich (1933-41) and professor of medical psychology at the University of Basel (1943). His personal experience, his continued psychotherapeutic practice, and his wide knowledge of history placed him in a unique position to comment on current events.

As early as 1918 he had begun to think that Germany held a special position in Europe; the Nazi revolution was, therefore, highly significant for him, and he delivered a number of hotly contested views that led to his being wrongly branded as a Nazi sympathizer. Jung lived to the age of 85.

5.5 RANK, OTTO

Otto Rank's original name was Otto Rosenfeld (1884 – 1939). An Austrian psychologist who extended psychoanalytic theory to the study of legend, myth, art, and creativity and who suggested that the basis of anxiety neurosis is a psychological trauma occurring during the birth of the individual.

Rank came from a poor family and attended trade school, working in a machine shop while trying to write at night. His reading of Sigmund Freud's *The Interpretation of Dreams* inspired him to write *Der Künstler* (1907; "The Artist"), an attempt to explain art by using psychoanalytic principles. This work brought him to the attention of Freud, who helped arrange his entry to the University of Vienna, from which he received his doctorate in philosophy in 1912.

While studying at the university, he legally adopted his pen name of Otto Rank and published two more works, *Der Mythus von der Geburt des Helden* (1909; *The Myth of the Birth of the Hero*) and *Das Inzest-Motiv in Dichtung und Sage* (1912; "The Incest Motif in Poetry and Saga"), in which he attempted to show how the Oedipus complex supplies abundant themes for poetry and myth.

Rank served as secretary to the Vienna Psychoanalytic Society and as editor of its minutes, and from 1912 to 1924 he edited the *Internationale Zeitschrift für Psychoanalyse* ("International Journal of Psychoanalysis"). In 1919 he founded a publishing house devoted to the publication of psychoanalytic works and directed it until 1924.

Publication of *Das Trauma der Geburt und seine Bedeutung für die Psychoanalyse* (1924; *The Trauma of Birth*) caused Rank's break with Freud and other members of the Vienna Psychoanalytic Society, which expelled him from its membership. The book, which argued that the transition from the womb to the outside world causes tremendous anxiety in the infant that may persist as anxiety neurosis into adulthood, was seen by many members of the Viennese society as conflicting with the concepts of psychoanalysis. Following the break, which became complete in the mid-1920s, Rank taught and practiced in the United States and Europe (chiefly Paris) for about 10 years, settling in New York City in 1936.

During the 1930s Rank developed a concept of the will as the guiding force in personality development. The will could be a positive force for controlling and using a person's instinctual drives, which were seen by Freud as the motivating factors in human behaviour. Thus, in Rank's view, resistance by a patient during psychoanalysis was a manifestation of this will and not inherently a negative factor; instead of wearing down such resistance, as a Freudian analyst would attempt, Rank would use it to direct self-discovery and development.

Rank's attempt to reduce all of psychology to a monolithic system based on the birth trauma is viewed as a serious departure from a scientific orientation. But his emphasis on personal growth and self-actualization and his application of psychoanalytic theory to the interpretation of art and myth have remained influential.

6. ANALYTICAL CONCEPTS

6.1 Psychoanalysis

In simple psychoanalytical terms these are the drives and impulses derived from the genetic background and concerned with the preservation and propagation of life. The ego, according to Freud, operates in conscious and preconscious levels of awareness. It is the portion of the personality concerned with the tasks of reality: perception, cognition, and executive actions. In the superego lie the individual's environmentally derived ideals and values and the mores of his family and society; the superego serves as a censor on the ego functions.

In the Freudian framework, conflicts among the three structures of the personality are repressed and lead to the arousal of anxiety. The person is protected from experiencing anxiety directly by the development of defence mechanisms, which are learned through family and cultural influences. These mechanisms become pathological when they inhibit pursuit of the satisfactions of living in a society. The existence of these patterns of adaptation or mechanisms of defence are quantitatively but not qualitatively different in the psychotic and neurotic states.

Freud held that the patient's emotional attachment to the analyst represented a transference of the patient's relationship to parents or important parental figures. Freud held that those strong feelings, unconsciously projected to the analyst, influenced the patient's capacity to make free associations. By objectively treating these responses and the resistances they evoked and by bringing the patient to analyze the origin of those feelings, Freud concluded that the analysis of the transference and the patient's resistance to its analysis were the keystones of psychoanalytic therapy.

Early schisms over such issues as the basic role that Freud ascribed to biological instinctual processes caused onetime associates Carl Jung, Otto Rank, and Alfred Adler to establish their own psychological theories. Most later controversies, however, were over details of Freudian theory or technique and did not lead to a complete departure from the parent system.

Other influential theorists have included Erik Erikson, Karen Horney, Erich Fromm, and Harry Stack Sullivan. At one time psychiatrists held a monopoly on psychoanalytic practice, but soon non-medical therapists also were admitted.

Later developments included work on the technique and theory of psychoanalysis of children. Freud's tripartite division of the mind into id, ego, and superego became progressively more elaborate, and problems of anxiety and female sexuality received increasing attention. Psychoanalysis also found many extra-clinical applications in other areas of social thought, particularly anthropology and sociology, and in literature and the arts.

6.2 Hysteria Disorder

Hysteria is a type of psychiatric disorder in which a wide variety of sensory, motor, or psychic disturbances may occur. It is traditionally classified as one of the psychoneuroses and is not dependent upon any known organic or structural pathology. The term is derived from the Greek *hystera,* meaning "uterus," and reflects the ancient notion that hysteria was a specifically female disorder resulting from disturbances in uterine functions.

Actually, hysterical symptoms may develop in either sex and may occur in children and elderly people, although they are observed most commonly in early adult life.

Hysteria, in its clinically pure form, seems to occur more often among the psychologically naive than among sophisticated persons. Hysterias tend to be more common among persons in the lower ranges of intelligence than among those in the higher ranges.

The incidence of hysteria appears to have been diminishing over the years in many areas of the world, probably because of cultural factors such as increasing psychological sophistication, diminishing sexual prudery and inhibition, and a less authoritarian family structure. Cases of classical hysteria, such as those frequently described by 19th-century clinicians, have become rare.

Most psychoneuroses encountered in actual clinical practice are apt to be "mixed" forms in which hysterical symptoms may be found interspersed with other varieties of neurotic disturbances. Isolated hysterical symptoms may also occur in conjunction with psychotic disorders.

The sensory and motor manifestations of hysteria take many forms and are designated conversion reactions because the underlying anxiety is assumed to have been "converted" into a physical symptom. Sensory disturbances may range from paresthesias ("peculiar" sensations) through hyperesthesias (hypersensitivity) to complete anesthesias (loss of sensation).

They may involve the total skin area or any fraction of it, but the disturbances generally do not follow any anatomic distribution of the nervous system. In medieval times in Europe and as late as the end of the 17th century, the finding of such discrete areas of anesthesia on the body of a person was considered proof that the person was a witch. Other hysterical sensory disturbances may encompass the special senses of vision, hearing, taste, or smell; or they may involve the experiencing of severe pain for which no organic cause can be determined.

Motor symptoms vary from complete paralysis to tremors, tics, contractures, or convulsions (hystero-epilepsy). In each instance neurological examination of the affected part of the body reveals an intact neuromuscular apparatus with normal reflexes and normal electrical activity and responses to electrical stimulation. Other motor disturbances that are at times hysterical in origin are loss of speech (aphonia), coughing, nausea, vomiting, or hiccuping.

Psychic symptoms may be equally varied and are usually classified under the broad heading of dissociative reactions. Attacks of amnesia, in which the person is unable to remember who he is or anything about himself, are among the more striking of these. Sleepwalking (somnambulism) is also considered to be a hysterical dissociative reaction, as are also the occasional dramatic cases of multiple personality.

6.3 Anxiety

Anxiety is a feeling of dread, fear, or apprehension, often with no clear justification. Anxiety is distinguished from true fear because the latter arises in response to a clear and actual danger, such as one affecting a person's physical safety.

Anxiety, by contrast, arises in response to apparently innocuous situations or is the product of subjective, internal emotional conflicts the causes of which may not be apparent to the person himself. Some anxiety inevitably arises in the course of daily life and is normal. But persistent, intense, chronic, or recurring anxiety not justified in response to real-life stresses is usually regarded as a sign of an emotional disorder.

When such an anxiety is unreasonably evoked by a specific situation or object, it is known as a phobia . A diffuse or persistent anxiety associated with no particular cause or mental concern is called general, or free-floating, anxiety.

There are many causes (and psychiatric explanations) for anxiety. Sigmund Freud viewed anxiety as the symptomatic expression of the inner emotional conflict caused when a person suppresses from conscious awareness experiences, feelings, or impulses that are too threatening or disturbing to live with.

Anxiety is also viewed as arising from threats to an individual's ego or self-esteem, as in the case of inadequate sexual or job performance. Behavioural psychologists view anxiety as an unfortunate learned response to frightening events in real life; the anxiety produced becomes attached to the surrounding circumstances associated with that event, so that those circumstances come to trigger anxiety in the person independently of any frightening event.

An anxiety disorder may develop where anxiety is insufficiently managed, characterized by a continuing or periodic state of anxiety or diffuse fear that is not restricted to definite situations or objects, and is generally classed as one of the psychoneuroses (neuroses).

The tension is frequently expressed in the form of insomnia, outbursts of irritability, agitation, palpitations of the heart, and fears of death or insanity. Fatigue is often experienced as a result of excessive effort expended in managing the distressing fear. Occasionally the anxiety is expressed in a more acute form and results in physiological concomitants such as nausea, diarrhoea, urinary frequency, suffocating sensations, dilated pupils, perspiration, and rapid breathing.

Similar symptoms occur in several physiological disorders and in normal situations of stress or fear, but they may be considered neurotic when they occur in the absence of any organic defect or pathology and in situations that most people handle with ease.

Other types of anxiety-related disorders include hypochondriasis, hysteria, obsessive-compulsive disorders, phobias, and schizophrenia.

6.4 Libido

The Libido concept was originated by Sigmund Freud to signify the instinctual physiological or psychic energy associated with sexual urges and, in his later writings, with all constructive human activity. In the latter sense of eros, or life instinct, libido was opposed by Thanatos, the death instinct and source of destructive urges; the interaction of the two produced all the variations of human activity. Freud considered psychiatric symptoms the result of misdirection or inadequate discharge of libido.

Carl Jung used the term in a more expansive sense, encompassing all life processes in all species. Later theories of motivation have substituted for libido such related terms as drive and tension.

6.5 Id

The Id, in Freudian psychoanalytic theory, is one of the three agencies of the human personality, along with the ego and superego. The oldest of these psychic realms in development, it contains the psychic content related to the primitive instincts of the body, notably sex and aggression, as well as all psychic material that is inherited and present at birth.

The id (Latin for "it") is oblivious of the external world and unaware of the passage of time. Devoid of organization, knowing neither logic nor reason, it has the ability to harbour acutely conflicting or mutually contradictory impulses side by side. It functions entirely according to the pleasure-pain principle, its impulses either seeking immediate fulfilment or settling for a compromise fulfilment.

The id supplies the energy for the development and continued functioning of conscious mental life, though the working processes of the id itself are completely unconscious in the adult (less unconscious in the child). In waking life it belies its content in slips of the tongue, wit, art, and other at least partly non-rational modes of expression. The primary methods for unmasking its content, according to Freud, are the analysis of dreams and free association.

Many psychoanalysts now consider the conception of an id overly simple, though still useful in drawing attention to the unconscious motivations and irrational impulses within even the most normal human being.

6.6 Ego

The Ego in psychoanalytic theory, that portion of the human personality which is experienced as the "self" or "I" and is in contact with the external world through perception. It is the part which remembers, evaluates, plans, and in other ways is responsive to and acts in the surrounding physical and social world. The ego coexists, in psychoanalytic theory, with the id and superego , as one of three agencies proposed by Sigmund Freud in attempting to describe the dynamics of the human mind.

The Ego (Greek and subsequently Latin for "I") comprises, in Freud's term, the executive functions of personality; it is the integrator between the outer and inner worlds, as well as between the id and the superego. The ego gives continuity and consistency to behaviour by providing a personal point of reference, which relates the events of the past (retained in memory) and actions of the present and of the future (represented in anticipation and imagination).

The ego is not coextensive with either the personality or the body, although body concepts form the core of early experiences of self. The ego, once developed, is capable of change throughout life, particularly under conditions of threat, illness, and changes in life circumstances.

6.7 Superego

The Superego according to the psychoanalytic theory of Sigmund Freud, latest developing of three agencies (with the id and ego;) of the human personality. The superego is the ethical component of the personality and provides the moral standards by which the ego operates. The superego's criticisms, prohibitions, and inhibitions form a person's conscience, and its positive aspirations and ideals represent one's idealized self-image, or "ego ideal."

The superego develops during the first five years of life in response to parental punishment and approval. This development occurs as a result of the child's internalization of his parents' moral standards, a process greatly aided by a tendency to identify with the parents.

The developing superego absorbs the traditions of the family and the surrounding society and serves to control aggressive or other socially unacceptable impulses. Violation of the superego's standards results in feelings of guilt or anxiety and a need to atone for one's actions.

The superego continues to develop into young adulthood as a person encounters other admired role models and copes with the rules and regulations of the larger society. *See also* Oedipus complex.

6.8 Oedipus

The Oedipus complex in psychoanalytic theory, is a desire for sexual involvement with the parent of the opposite sex and a concomitant sense of rivalry with the parent of the same sex; a crucial stage in the normal developmental process. Sigmund Freud introduced the concept in his *Interpretation of Dreams* (1899).

The term derives from the Theban hero Oedipus of Greek legend, who unknowingly slew his father and married his mother; its female analogue, the Electra complex, is named for another mythological figure, who helped slay her mother.

Freud attributed the Oedipus complex to children of about the ages three to five. He said the stage usually ended when the child identified with the parent of the same sex and repressed its sexual instincts. If previous relationships with the parents were relatively loving and non-traumatic, and if parental attitudes were neither excessively prohibitive nor excessively stimulating, the stage is passed through harmoniously.

In the presence of trauma, however, there occurs an "infantile neurosis" that is an important forerunner of similar reactions during the child's adult life. The superego, the moral factor that dominates the conscious adult mind, also has its origin in the process of overcoming the Oedipus complex. Freud considered the reactions against the Oedipus complex the most important social achievements of the human mind.

6.9 Psychoanalytic Dreams

Among Freud's earliest writings was *The Interpretation of Dreams* (1899). His insistence that dreams are "the royal road to the unconscious" continued from it down to his last published statement on dreams, printed about a year before he died. Freud held that dreams reflect waking experience; he offered a theoretical explanation for their bizarre nature, invented a system for their interpretation, and elaborated on their curative potential.

Freud theorised that thinking during sleep tends to be primitive and regressive and that the effects of forgetting (repression) are reduced. Repressed wishes, particularly those associated with sex and hostility, were said to be released in dreams when the inhibitory demands of wakefulness diminished.

The content of the dream was said to derive from such stimuli as urinary pressure in the bladder, traces of experiences from the previous day (day residues), and associated infantile memories. The specific dream details were called their manifest content; the presumably repressed wishes being expressed were called the latent content.

Freud suggested that the dreamer kept himself from waking and avoided unpleasant awareness of repressed wishes by disguising them as bizarre manifest content in an effort called dream-work. He held that impulses one fails to satisfy when awake are expressed in dreams as sensory images and scenes. In dreaming, Freud believed:

All of the linguistic instruments . . . of subtle thought are dropped . . . and abstract terms are taken back to the concrete . . . . The copious employment of symbols . . . for representing certain objects and processes is in harmony (with) the regression of the mental apparatus and the demands of censorship.

Freud theorised that one aspect of manifest content could come to represent a number of latent elements (and vice versa) through a process called condensation. Further displacement of emotional attitudes toward one object or person theoretically could be displaced in dreaming to another object or person or not appear in the dream at all. Freud further observed a process called secondary elaboration, which occurs when people wake and try to remember dreams. They may recall inaccurately in a process of elaboration and rationalization and provide "the dream, a smooth facade, (or by omission) display rents and cracks." This waking activity he called secondary revision.

In seeking the latent meaning of a dream, Freud advised the individual to associate freely about it. From listening to the associations, the analyst was supposed to determine what the dream represented, in part through an understanding of the personal needs of the dreamer.

6.10 Disagreement

Carl Jung (1875-1961) disagreed with Freud's view of dreams as being complementary to waking mental life with respect to specific instinctual impulses. Jung felt that dreams are instead compensatory, that they balance whatever elements of character are underrepresented in the way people are living their lives. Dreaming, to Jung, represents a continuous 24-hour flow of mental activity that surfaces in sleep when conditions are right, but which affects waking life when a person's behaviour denies important elements of his true personality.

Thus, dreams are constructed not to conceal or disguise forbidden wishes but to bring the under-attended areas to attention. This function is carried out unconsciously in sleep when people are living well-balanced lives. If this is not the case there may be first bad moods, then symptoms in waking. Then and only then do dreams need to be interpreted. This is best done not with a single dream and multiple free associations but with a series of dreams so that the repetitive elements become apparent.

6.11 Unconscious

The unconscious, also called SUBCONSCIOUS, is the complex of mental activities within an individual that proceed without his awareness. Sigmund Freud, the founder of psychoanalysis, stated that such unconscious processes may affect a person's behaviour even though he cannot report on them. Freud and his followers felt that dreams and slips of the tongue were really concealed examples of unconscious content too threatening to be confronted directly.

Some theorists (*e.g.*, the early experimental psychologist Wilhelm Wundt) denied the role of unconscious processes, defining psychology as the study of conscious states. Yet, the existence of unconscious mental activities seems well established and continues to be an important concept in modern psychiatry.

6.12 Levels of Consciousness

Freud distinguished among different levels of consciousness. Activities within the immediate field of awareness he termed conscious; *e.g.*, reading this article is a conscious activity. The retention of data easily brought to awareness is a preconscious activity; for example, one may not be thinking (conscious) of his address but readily recalls it when asked. Data that cannot be recalled with effort at a specific time but that later may be remembered are retained on an unconscious level. For example, under ordinary conditions a person may be unconscious of ever having been locked in a closet as a child; yet under hypnosis he may recall the experience vividly.

Because one's experiences cannot be observed directly by another (as one cannot feel another's headache), efforts to study these levels of awareness objectively are based on inference; *i.e.*, at most, the investigator can say only that another individual behaves *as if* he were unconscious or *as if* he were conscious.

Efforts to interpret the origin and significance of unconscious activities lean heavily on psychoanalytic theory, developed by Freud and his followers. For example, the origin of many neurotic symptoms is held to depend on conflicts that have been removed from consciousness through a process called repression.

As knowledge of psychophysiological function grows, many psychoanalytic ideas are seen to be related to activities of the central nervous system. That the physiological foundation of memory may rest in chemical changes occurring within brain cells has been inferred from clinical observations that:

(1) Direct stimulation of the surface of the brain (the cortex) while the patient is conscious on the operating table during surgery has the effect of bringing long-forgotten (unconscious) experiences back to awareness;

(2) Removal of specific parts of the brain seems to abolish the retention of specific experiences in memory;

(3) The general probability of bringing unconscious or preconscious data to awareness is enhanced by direct electrical stimulation of a portion of the brain structure called the reticular formation, or the reticular activating system.

Also, according to what is called brain blood-shift theory, the transition from unconscious to conscious activities is mediated by localized changes in the blood supply to different parts of the brain. These bio-psychological explorations have shed new light on the validity of psychoanalytic ideas about the unconscious. *See also* psychoanalysis.

6.13 Psychotherapies

Psychotherapy implies the treatment of mental discomfort, dysfunction, or disease by psychological means by a trained therapist who adheres to a particular theory of both symptom causation and symptom relief.

The American psychotherapist Jerome Frank has classified psychotherapies into religio-magical and empirico-scientific forms. The former depend on the shared beliefs of the therapist and client in magic, spirits, or other supernatural processes or powers. This article is concerned, however, with the latter forms of psychotherapy, which have been developed by modern medicine and which are carried out by a member of one of the mental health professions such as a psychiatrist or a clinical psychologist.

It is usual to contrast two main forms of psychotherapy, dynamic and behavioural. They are conceptually different; behaviour therapy concentrates on alleviating a patient's overt symptoms, which are attributed to faulty learning, while dynamic therapy concentrates on understanding the meaning of symptoms and understanding the emotional conflicts within the patient that may be causing those symptoms. In their pure forms the two approaches are very different, but in practice many therapists use elements of both.

6.14 Psychiatry

Psychiatry is the branch of medicine that is concerned with the diagnosis, treatment, and prevention of mental disorders.

The term *psychiatry* is derived from two Greek words meaning "mind healing." Until the 18th century, mental illness or disorder was most often seen as demonic possession, but it gradually came to be considered as a sickness requiring treatment.

Many judge that modern psychiatry was born with the efforts of Philippe Pinel in France and J. Connolly in England, who both advocated a more humane approach to mental illness. By the 19th century, research, classification, and treatment of disorders had gained momentum. Psychotherapy evolved from its origins in spiritual healing.

The psychoanalytic theory of Sigmund Freud and his followers dominated the field for many years and did not receive a serious theoretical challenge until behaviour therapy and therapies deriving from humanistic psychology were developed in the mid-20th century. Insight therapies such as psychoanalysis, which pursue greater awareness of the patient's internal conflicts, continue to be dominant in psychiatric practice.

The trained psychiatrist, who has completed medical school and a psychiatric residency, commonly employs medical treatments in addition to psychotherapy. Lobotomy, or leucotomy, whereby nerve fibres running to the front of the brain are severed, is today used only in severe cases and has generally lost favour as a treatment.

Shock therapy (also called electroshock, or electroconvulsive, therapy) continues to be used for severe depressions and certain forms of psychosis. The medical technique that is by far the most widely used is drug therapy. The advent in the 1950s of psychotropic (mind-altering) drugs revolutionized treatment of the mental patient. Like the other medical techniques, drug therapy has sometimes been abused in pursuit of patient "management"; used properly, however, it can enhance a patient's outlook for recovery and return to the community.

The contemporary psychiatrist frequently functions as a member of a mental-health team that includes clinical psychologists and social workers. As the therapeutic roles of these three professionals are not necessarily clearly delineated, an uneasy balance in orientation and division of skills may exist.

6.15 Mental Hygiene

Since the founding of the United Nations the concepts of mental health and hygiene have achieved international acceptance. As defined in the 1946 constitution of the World Health Organization, "health is a state of complete physical, mental, and social well-being, and not merely the absence of disease or infirmity."

The term mental health represents a variety of human aspirations: rehabilitation of the mentally disturbed, prevention of mental disorder, reduction of tension in a stressful world, and attainment of a state of well-being in which the individual functions at a level consistent with his or her mental potential. As noted by the World Federation for Mental Health, the concept of optimum mental health refers not to an absolute or ideal state but to the best possible state insofar as circumstances are alterable.

Mental health is regarded as a condition of the individual, relative to the capacities and social-environmental context of that person. Mental hygiene includes all measures taken to promote and to preserve mental health. Community mental health refers to the extent to which the organization and functioning of the community determines, or is conducive to, the mental health of its members.

Throughout the ages the mentally disturbed have been viewed with a mixture of fear and revulsion. Their fate generally has been one of rejection, neglect, and ill treatment. Though in ancient medical writings there are references to mental disturbance that display views very similar to modern humane attitudes, interspersed in the same literature are instances of socially sanctioned cruelty based upon the belief that mental disorders have supernatural origins such as demonic possession. Even reformers sometimes used harsh methods of treatment; for example, the 18th-century American physician Benjamin Rush endorsed the practice of restraining mental patients with his notorious "tranquilising chair."

6.16 Child Development

This term refers to the growth of perceptual, emotional, intellectual, and behavioural capabilities and functioning during childhood. The term childhood denotes that period in the human lifespan from the acquisition of language at one or two years to the onset of adolescence at 12 or 13 years.

The end of infancy and the onset of childhood are marked by the emergence of speech at one to two years of age. Children make enormous progress in language acquisition in their second year and demonstrate a continually growing vocabulary, an increasing use of words in combinations, and a dawning understanding of the rules of grammar and syntax. By their third year children tend to use sentences containing five or even six words, and by the fourth year they can converse in adult-like sentences. Five- and six-year-olds demonstrate a mastery of complex rules of grammar and meaning.

Early childhood (two to seven years) is also the time in which children learn to use symbolic thought and language to manipulate their environment. They learn to perform various mental operations using symbols, concepts, and ideas to transform information they gather about the world around them. The beginnings of logic, involving the classification of ideas and an understanding of time and number, emerge in later childhood (7 to 12 years).

Children's memory capacity also advances continually during childhood and underpins many other cognitive advances they make at that time. As both short-term and long-term memory improves, children demonstrate an increasing speed of recall and can search their memory for information more quickly and efficiently.

Young children's growing awareness of their own emotional states, characteristics, and abilities lead to empathy--*i.e.*, the ability to appreciate the feelings and perspectives of others. Empathy and other forms of social awareness are in turn important in the development of a moral sense. The basis of morality in children may be said to progress from a simple fear of punishment and pain to a concern for maintaining the approval of one's parents.

Another important aspect of children's emotional development is the formation of their self-concept, or identity--*i.e.*, their sense of who they are and what their relation to other people is. Sex-role identity, based on gender, is probably the most important category of self-awareness and usually appears by the age of three.

The onset of the physical and emotional changes of puberty and the acquisition of the logical processes of adults mark the end of childhood and the start of adolescence.

6.17 Child Psychology

Child psychology, also called CHILD DEVELOPMENT, is the study of the psychological processes of children, specifically, how these processes differ from those of adults, how they develop from birth to the end of adolescence, and how and why they differ from one child to the next. The topic is sometimes subsumed with infancy, adulthood, and aging under the category of developmental psychology.

As a scientific discipline with a firm empirical basis, child study is of comparatively recent origin. It was initiated in 1840, when Charles Darwin began a record of the growth and development of one of his own children, collecting the data much as if he were studying some strange species. A similar, more elaborate study was published by the German psycho-physiologist W.T. Preyer (*Die Seele des Kindes* [1882; *The Mind of the Child*]) that set the fashion for a series of others. In 1891 the American educational psychologist G. Stanley Hall established a periodical, the *Pedagogical Seminary,* devoted to child psychology and pedagogy. During the early 20th century, the development of intelligence tests and the establishment of child guidance clinics further defined the field of child psychology.

A number of notable 20th-century psychologists--among them Sigmund Freud, Melanie Klein, and Freud's daughter, Anna Freud--dealt with child development chiefly from the psychoanalytic point of view. Perhaps the greatest direct influence on modern child psychology was Jean Piaget of Switzerland. By means of direct observation and interaction, Piaget developed a theory based on the systematic study of the acquisition of understanding in children. He described the various stages of learning in childhood and characterized the child's perception of himself and the world at each stage.

The data of child psychology are gathered from a variety of sources. Observations by relatives, teachers, and other adults, as well as the psychologist's direct observation of and interviews with a child (or children), provide a significant amount of material. In some cases a one-way window or mirror is used so that children are free to interact with their environment or others without awareness that they are being watched. Projective tests, personality and intelligence tests, and experimental methods have also proved useful in understanding child development.

Despite attempts to unify the various theories of child development, the field remains dynamic, developing as human understanding of physiology and psychology changes.

7. PSYCHOLOGY AND BRANCHES

7.1 Scientific Discipline

The generalised term of psychology refers to the studies which attempt to create a scientific discipline that studies mental processes and behaviour in humans and other animals.

Psychology, therefore, can be considered to be the science of individual or group behaviour. The word psychology literally means "study of the mind"; the issue of the relationship of mind and body is pervasive in psychology, owing to its derivation from the fields of philosophy and physiology. Psychology is intimately related to the biological and social sciences.

The broad reach of psychology sometimes gives it the appearance of disunity and promotes the lack of a universally accepted theoretical structure. Some of the divisions within psychology are applied fields, while others are more experimental in nature.

The various applied fields include clinical; counselling; industrial, engineering, or personnel; consumer; and environmental. The most important of these specialties, clinical psychology, is concerned with the diagnosis and treatment of mental disorders. Industrial psychology is used in employee selection and related contexts in business and industry.

The broad field known as experimental psychology includes specializations in child, educational, social, developmental, physiological, and comparative psychology. Of these, child psychology applies psychological theory and research methods to children; educational psychology is concerned with learning processes and problems associated with the teaching of students; social psychology is concerned with group dynamics and other aspects of human behaviour in its social and cultural setting; and comparative psychology deals with behaviour as it differs from one species of animal to another.

The issues studied by psychologists cover a wide spectrum, comprising learning, cognition, intelligence, motivation, emotion, perception, personality, mental disorders, and the study of the extent to which individual differences are inherited or are shaped environmentally, known as behaviour genetics.

7.2 History of psychology

The history of psychology is the history of thought about human consciousness and conduct. Psychological theory has its roots in ancient Greek philosophy and has been fed from streams such as epistemology (the philosophy of knowing), metaphysics, religion, and Oriental philosophy.

Over the centuries psychology and physiology became increasingly separated. A split developed between the essentially phenomenological (experiential) and mechanistic (physiological) conceptions of psychology. In general, through the end of the the 19th century the British and German traditions were phenomenological, while the French and American were mechanistic.

The history of psychology from the 19th century may be viewed as a debate between schools of systematic thought concerning the mind, such as associationism, structuralism, and functionalism; or alternatively, as a history of experimentation and research in various areas.

Twentieth-century psychology began with structuralism, which employed the method of introspection to describe mental events. It then evolved into psychoanalysis, a derivative of psychiatric tradition, and produced behaviourism and Gestalt psychology, which were reactions against structuralism. Humanistic psychology represented a rebellion against the reductionist and deterministic leanings of earlier schools.

By World War II, "schools" of psychology had largely faded away, leaving a common pool of psychological knowledge to which theoreticians, researchers, experimenters, and clinicians all contributed. Biopsychology, a study combining psychology and physiology, grew in conjunction with these developments.

7.3 Logic, Philosophy of Psychology

Although the "laws of thought" studied in logic are not the empirical generalizations of a psychologist, they can serve as a conceptual framework for psychological theorizing. Probably the best known recent example of such theorizing is the large-scale attempt made in the mid-20th century by Jean Piaget, a Swiss psychologist, to characterize the developmental stages of a child's thought by reference to the logical structures that he can master.

Elsewhere in psychology, logic is employed mostly as an ingredient of various models using mathematical ideas or ideas drawn from such areas as automata or information theory. Large-scale direct uses are rare, however, partly because of the problems mentioned above in the section on logic and information.

7.4 Aristotle and Works on Psychology

The relation between the active principle and the passive continuum (or between form and matter) that is operational in sentient and intellectual life is examined in *On the Soul*. After exploring the concept and the conditions of life, Aristotle relates the function of matter and form (body and soul) in human life to all of life's biological and psychological phenomena while rejecting Platonic transcendentalist and pre-Socratic materialist theories on the nature of the soul.

The soul, as the form of the organic body, consists of an ordered set of faculties; these are, in hierarchical order, the nutritive, the perceptual, and the intellectual faculties. The nutritive faculty is common to all living things and is responsible for growth and nutrition; the perceptual faculty is common to all animals and is responsible for, among other things, sight, hearing, smell, and locomotion; and the intellectual faculty is peculiar to humans.

Aristotle gives detailed accounts of the modes of perception (in addition to the five senses and their objects he postulates the existence of a "common sense" that unites their deliverances) and a notoriously difficult account of thought (which distinguishes an "active" from a "passive" intellect). The work also contains a discussion of animal movement and of its preconditions--of imagination and of desire.

7.5 Sense and Sensible

In the *Parva Naturalia*, the medieval designation for a collection of short treatises on natural functions, the argument of *On the Soul* is supplemented by a sequence of treatises on sense and the sensible, memory and reminiscence, sleeping and waking, prophecy in sleep, the length and brevity of life, youth and old age, life and death, and respiration.

7.6 Interest in Psychology

In 1872 William James was appointed instructor in physiology at Harvard College, in which capacity he served until 1876. But he could not be diverted from his ruling passion, and the step from teaching physiology to teaching psychology--not the traditional "mental science" but physiological psychology--was as inevitable as it was revolutionary.

It meant a challenge to the vested interests of the mind, mainly theological that were entrenched in the colleges and universities of the United States; and it meant a definite break with what Santayana called "the genteel tradition." Psychology ceased to be mental philosophy and became a laboratory science. Philosophy ceased to be an exercise in the grammar of assent and became an adventure in methodological invention and metaphysical discovery.

With his marriage in 1878, to Alice H. Gibbens of Cambridge, Mass., a new life began for James. The old neurasthenia practically disappeared. He went at his tasks with a zest and an energy of which his earlier record had given no hint. It was as if some deeper level of his being had been tapped: his life as an originative thinker began in earnest. He contracted to produce a textbook of psychology by 1880. But the work grew under his hand, and when it finally appeared in 1890, as *The Principles of Psychology,* it was not a textbook but a monumental work in two great volumes, from which the textbook was condensed two years later.

The *Principles,* which was recognized at once as both definitive and innovating in its field, established the functional point of view in psychology. It assimilated mental science to the biological disciplines and treated thinking and knowledge as instruments in the struggle to live. At one and the same time it made the fullest use of principles of psychophysics (the study of the effect of physical processes upon the mental processes of an organism) and defended, without embracing, free will.

7.7 Comparative Psychology

Comparative psychology is the branch which deals with similarities and differences in animal (including human) behaviour. It has important applications in fields such as medicine, ecology, and animal training. With the rise of an experimental comparative psychology in the latter half of the 19th century and its rapid growth during the 20th, the study of lower animals has cast increasing light on human psychology in such areas as the development of individual behaviour, motivation, the nature and methods of learning, effects of drugs, and localization of brain function.

Other animals are easier to obtain in numbers and can be better controlled under experimental conditions than can human subjects, and much can be learned about humans from lower animals. Comparative psychologists have been careful, however, to avoid anthropomorphizing the behaviour of animals; that is, to avoid ascribing to animals human attributes and motivations when their behaviours can be explained by simpler theories. This principle is known as Lloyd Morgan's canon, named after a British pioneer in comparative psychology.

The tendency to endow lower animals with human capacities always has been strong. In recorded history, two different views have developed concerning human beings' relation to the lower animals. One, termed for convenience the man-brute view, stresses differences often to the point of denying similarities altogether and derives from the traditional religious accounts of the separate creations of humans and animals; the other, the evolutionary view, stresses both similarities and differences. Aristotle formalized the man-brute view, attributing a rational faculty to humans alone, lesser faculties to the animals. The modern scientific view, on the other hand, considers humans to be highly evolved animals; evidence indicates that continuity in the evolution of organisms provides a basis for essential psychological similarities and differences between lower and higher animals, including humans.

7.8 Behaviourism

JOHN BROADUS WATSON is the American psychologist who codified and publicised behaviourism, an approach to psychology that, in his view, was restricted to the objective, experimental study of the relations between environmental events and human behaviour. Watsonian behaviourism became the dominant psychology in the United States during the 1920s and '30s.

Watson received his Ph.D. in psychology from the University of Chicago (1903), where he then taught. In 1908 he became professor of psychology at Johns Hopkins University, Baltimore, Md., and immediately established a laboratory for research in comparative, or animal, psychology. About this time he articulated his first statements on behaviourist psychology, and in the epoch-making article "Psychology as a Behaviorist Views It" (1913) he asserted that psychology is the science of human behaviour, which, like animal behaviour, should be studied under exacting laboratory conditions.

His first major work, *Behavior: An Introduction to Comparative Psychology*, was published in 1914. In it he argued forcefully for the use of animal subjects in psychological study and described instinct as a series of reflexes that are activated by heredity. He also promoted conditioned responses as the ideal experimental tool. In 1918 Watson ventured into the relatively unexplored field of infant study. In one of his classic experiments, he conditioned fear of white rats and other furry objects in an 11-month old boy.

The definitive statement of Watson's position appears in another major work, *Psychology from the Standpoint of a Behaviorist* (1919), in which he sought to extend the principles and methods of comparative psychology to the study of human beings and staunchly advocated the use of conditioning in research. His association with professional psychology ended abruptly. In 1920, in the wake of sensational publicity surrounding his divorce from his first wife, Watson resigned from Johns Hopkins.

Watson entered the advertising business in 1921. His book *Behaviourism* (1925), for the general reader, is credited with interesting many in entering professional psychology. Following

Psychological Care of Infant and Child (1928) and his revision (1930) of *Behaviourism,* he devoted himself exclusively to business until his retirement (1946).

7.9 Applied Psychology

Applied psychology includes the use of the findings and methods of scientific psychology in solving practical problems of human and animal behaviour and experience. A more precise definition is impossible because the activities of applied psychology range from laboratory experimentation through field studies of specific utility to direct services to troubled persons.

The same intellectual streams whose confluence produced psychology as an independent discipline in the latter part of the 19th century led to the later development of an applied psychology. Francis Galton's publication in 1883 of *Inquiries Into Human Faculty* foreshadowed the measurement of individual psychological differences. In 1896 Lightner Witmer established at the University of Pennsylvania, Philadelphia, a clinic that was a forerunner of clinical psychology.

Intelligence testing began with the work of Alfred Binet and Théodore Simon in the Paris schools. Group testing, legal problems, industrial efficiency, motivation, and delinquency were among other early areas of application. At the Carnegie Institute of Technology, Pittsburgh, a division of applied psychology was established as a teaching and research department in 1915. The *Journal of Applied Psychology* appeared in 1917 along with the first applied-psychology text, by H.L. Hollingsworth and A.T. Poffenberger. World Wars I and II fostered work on vocational testing, teaching methods, evaluation of attitudes and morale, performance under stress, propaganda and psychological warfare, rehabilitation, and counselling.

After World War II many of the trends in applied psychology were accentuated by the demands of the space age. Educational psychologists applied themselves to the task of early identification and discovery of talented persons, since it was recognized that trained intelligence is an important national resource. Such activities were linked with the work of counselling psychologists, who sought to help persons clarify and attain their educational, vocational, and personal goals.

Concern for the optimum utilization of human resources also increased the importance of industrial and personnel psychology in business and industrial organizations. The aviation industry and the various space agencies and organizations were important in the rapid development of the field of engineering psychology; as machines and engineering systems grew in complexity, it was necessary to study man-machine relationships. In response to society's concern for treatment of the mentally ill and for preventive measures against mental illness, clinical psychology showed the greatest absolute growth rate within psychology.

Psychologists studied the application of automation, and in the developing countries they helped with the problems of rapid industrialization and manpower planning.

Regardless of the applied psychologist's professional focus, his job description is likely to overlap with those of other areas. The applied psychologist may or may not engage in original research and/or teach. In addition to drawing on experimental findings gleaned from psychological research, the applied psychologist utilizes information from many disciplines. The scope of the field is continually broadening as new types of problems (*e.g.,* technological) arise. Other branches of applied psychology include consumer, school, and community psychology. Prevention and treatment of emotional problems in naturalistic settings (*i.e.,* the community) have received a great deal of attention, as have medically related questions (*e.g.,* sports psychology and the psychology of chronic illness).

Psychometrics, or the measurement and evaluation of psychological variables such as personality, aptitude, or performance, is an integral part of applied-psychology fields. For example, the clinical psychologist may be interested in measuring the traits of aggressiveness or obsessiveness; the counselling psychologist, areas of career interest or aptitude; the industrial psychologist, work effectiveness under certain conditions of lighting or office design; or the community psychologist, psychological effects of living near a nuclear power plant or radioactive waste disposal site. *See also* clinical psychology; counselling; educational psychology; industrial psychology.

7.10 Experimental Psychology

Experimental psychology is a method of studying psychological problems; the term generally connotes all areas of psychology that use the experimental method. The experimental method in psychology is an attempt to account for the activities of animals (including humans) and the functional organization of their mental processes by manipulating variables that may give rise to behaviour; it is primarily concerned with discovering laws that describe manipulable relationships.

The areas of study in psychology that lean heavily on the experimental method include those of sensation and perception, learning and memory, motivation, and physiological psychology. There are experimental branches in most areas, however, including child psychology, clinical psychology, educational psychology, social psychology, and even parapsychology.

Usually the experimental psychologist deals with normal, intact organisms; but in physiological psychology, studies are often conducted with organisms modified by surgery, radiation, drug treatment, or long-standing deprivations of various kinds; or with organisms who naturally present organic abnormalities or emotional disorders.

7.11 Industrial Psychology

Industrial psychology, also called Occupational Psychology, covers the application of the concepts and methods of experimental, clinical, and social psychology to industry. The primary concern of industrial psychology is with the basic relations in industry between worker and machine and the organization.

Industrial psychology was first developed in the United States in the early 1900s and has come to be applied, usually through personnel and office administrations, to industrial management in industries in other parts of the world, particularly in those countries where a concern has developed for the systematic and scientific examination of the problems of workers in industry.

7.12 Social Psychology

Social psychology is a distinct discipline originated in the 19th century, although its outlines were perhaps somewhat less clear than was true of the other social sciences. The close relation of the human mind to the social order, its dependence upon education and other forms of socialization, was well known in the 18th century.

In the 19th century, however, an ever more systematic discipline came into being to uncover the social and cultural roots of human psychology and also the several types of "collective mind" that analysis of different cultures and societies in the world might reveal. In Germany, Moritz Lazarus and Wilhelm Wundt sought to fuse the study of psychological phenomena with analyses of whole cultures. Folk psychology, as it was called, did not, however, last very long in scientific esteem.

Much more esteemed, and closer to 20th-century conceptions of social psychology, were the works of such men as Gabriel Tarde, Gustave Le Bon, Lucien Lévy-Bruhl, and Émile Durkheim in France and Georg Simmel in Germany (all of whom also wrote in the early 20th century).

Here, in concrete, often highly empirical studies of small groups, associations, crowds, and other aggregates (rather than in the main line of psychology during the century, which tended to be sheer philosophy at one extreme and a variant of physiology at the other) are to be found the real beginnings of social psychology.

Although the point of departure in each of the studies was the nature of association, they dealt, in one degree or other, with the internal processes of psychosocial interaction, the operation of attitudes and judgments, and the social basis of personality and thought--in short, with those phenomena that would, in the 20th century, be the substance of social psychology as a formal discipline.

7.13 Developmental Psychology

Developmental psychology, also called LIFE-SPAN PSYCHOLOGY, is the branch of psychology concerned with the changes in cognitive, motivational, psychophysiological, and social functioning that occur throughout the human life span. During the 19th and early 20th centuries, developmental psychologists were concerned primarily with child psychology. In the 1950s, however, they became interested in the relationship between personality variables and child rearing, and the behavioural theories of B.F. Skinner and the cognitive theories of Jean Piaget were concerned with the growth and development of children through adolescence.

At the same time, the German psychologist Erik Erikson insisted that there are meaningful stages of adult psychology that have to be considered in addition to child development. Psychologists also began to consider the processes that underlie the development of behaviour in the total person from birth to death, including various aspects of the physical-chemical environment that can affect the individual during the intrauterine period and at birth.

By the latter part of the 20th century, developmental psychologists had become interested in many broad issues dealing with the psychological process throughout life, including the relation of heredity and environment, continuity and discontinuity in development, and behavioural and cognitive elements in the development of the total person.

7.14 Educational Psychology

Educational psychology is the theoretical and research branch of modern psychology, concerned with the learning processes and psychological problems associated with the teaching and training of students. The educational psychologist studies the cognitive development of students and the various factors involved in learning, including aptitude and learning measurement, the creative process, and the motivational forces that influence dynamics between students and teachers.

Educational psychology is a partly experimental and partly applied branch of psychology, concerned with the optimization of learning. It differs from school psychology, which is an applied field that deals largely with the problems in elementary and secondary school systems.

Educational psychology traces its origins to the experimental and empirical work on association and sensory activity by the English scientist and founder of eugenics, Sir Francis Galton (1822-1911), and the American psychologist G. Stanley Hall (1844-1924), who wrote *The Contents of Children's Minds* (1883).

The major leader in the field of educational psychology, however, was the American educator and psychologist Edward Lee Thorndike (1874-1949), who designed methods to measure and test children's intelligence and their ability to learn. Thorndike proposed the transfer-of-training theory, which states that "what is learned in one sphere of activity transfers' to another sphere only when the two spheres share common elements.' "

7.15 Psychopathology

Psychopathology, also called Abnormal Psychology, is the study of mental disorders and unusual or maladaptive behaviours. An understanding of the genesis of mental disorders is critical to mental health professionals in psychiatry, psychology, and social work. One controversial issue in psychopathology is the distinction between dysfunctional, or aberrant, and merely idiosyncratic behaviours.

7.16 Social Psychology

Social psychology is the scientific study of the behaviour of individuals in their social and cultural setting. Although the term may be taken to include the social activity of laboratory animals or those in the wild, the emphasis here is on human social behaviour.

Once a relatively speculative, intuitive enterprise, social psychology has become an active form of empirical investigation, the volume of research literature having risen rapidly after about 1925. Social psychologists now have a substantial volume of observation data covering a range of topics; the evidence remains loosely coordinated, however, and the field is beset by many different theories and conceptual schemes.

Early impetus in research came from the United States, and much work in other countries has followed U.S. tradition, though independent research efforts are being made elsewhere in the world. Social psychology is being actively pursued in the United Kingdom, Canada, Australia, Germany, The Netherlands, France,

Adler's Individual Psychology and Related Methods

Belgium, Scandinavia, Japan, and Russia. Most social psychologists are members of university departments of psychology; others are in departments of sociology or work in such applied settings as industry and government.

Much research in social psychology has consisted of laboratory experiments on social behaviour, but this approach has been criticized in recent years as being too stultifying, artificial, and unrealistic. Much of the conceptual background of research in social psychology derives from other fields of psychology. While learning theory and psychoanalysis were once most influential, cognitive and linguistic approaches to research have become more popular; sociological contributions also have been influential.

Social psychologists are employed, or used as consultants, in setting up the social organization of businesses and psychiatric communities; some work to reduce racial conflict, to design mass communications (*e.g.*, advertising), and to advise on child rearing. They have helped in the treatment of mental patients and in the rehabilitation of convicts. Fundamental research in social psychology has been brought to the attention of the public through popular books and in the periodical press.

7.17 Clinical Psychology

Clinical psychology is a branch of psychology concerned with the practical application of research findings and methodologies in the diagnosis and treatment of mental disorders.

Clinical psychologists classify their basic activities under three main headings: assessment (including diagnosis), treatment, and research. In assessment, clinical psychologists give and interpret psychological tests, either for the purpose of evaluating individuals' relative intelligence or other capabilities or for the purpose of eliciting mental characteristics that will aid in diagnosing a particular mental disorder. The interview, in which the psychologist observes, questions, and interacts with a patient, is another standard tool of diagnosis.

For purposes of treatment, the clinical psychologist may use any of several types of psychotherapy, and recently the tendency has been toward an eclectic approach, using a combination of techniques suited to the client. Clinical psychologists may specialize in behaviour therapy, group therapy, family therapy, or psychoanalysis, among others.

Research is an important field for some clinical psychologists because of their training in the use of experimental studies and statistical procedures. Clinical psychologists are thus often crucial participants in research projects bearing on mental-health care.

Clinical psychologists perform their services in hospitals, clinics, or in private practice, while others work with the mentally or physically handicapped, prison inmates, drug and alcohol abusers, or geriatric patients. In some clinical settings, a clinical psychologist works in tandem with a psychiatrist and a social worker and is responsible for conducting the team's research. Clinical psychologists are also employed in industry, where some specialize in services to emotionally disturbed employees and others in services for managerial officials. Other clinical psychologists serve the courts in assessing defendants or potential parolees, and some are employed by the armed forces to evaluate or treat service personnel.

The training of clinical psychologists usually includes the university-level study of general psychology and some clinical experience and amounts to 5-7 years of higher education in all. Because they have not earned a medical degree, clinical psychologists cannot prescribe medications for patients.

8. ADLERIAN APPROACH

8.1 Holism

Central to the Adlerian approach is to see the personality as a whole and not as the mere net result of component forces. Adlerians adopt a radical stance that cuts across the nature-nurture debate by seeing the developing individual at work in creating the personality in response to the demands of nature and nurture but not absolutely determined by them.

The self-created personality operates subjectively and idiosyncratically. The individual is endowed with a striving both for self-development and social meaning - what Adler himself called "the concept of social usefulness and the general well-being of humanity"- expressed in a sense of belonging, usefulness and contribution, and even cosmic consciousness.

8.2 Compensation

Neurosis and other pathological states reveal the safe-guarding or defensive mechanism of the individual who believes him- or herself to be unequal to the demands of life, in a struggle to compensate for a felt weakness, physical or psychological.

In "normal" development the child has experienced encouragement and accepts that his or her problems can be overcome in time by an investment of patient persistence and cooperation with others. The "normal" person feels a full member of life, and has "the courage to be imperfect".

In less fortunate circumstances, the child, trapped within a sense of inferiority, compensates, perhaps in grandiose fashion - by striving, consciously and unconsciously, to overcome and solve the problems of life. A high level of compensation produces subsequent psychological difficulties.

8.3 Withdrawal

In cases of discouragement the individual, feeling unable to unfold a real and socially valid development, erects a fantasy of superiority - what Adler termed "an attempt at a planned final compensation and a (secret) life plan" - in some backwater of life, which offers seclusion and shelter from the threat of failure and annihilation of personal prestige.

This fictional world, sustained by the need to safeguard an anxious ego, by private logic at variance with reason or common sense, by a schema of apperception which interprets and filters and suppresses the real-world data, is a fragile bubble waiting to be burst by mounting tension within and by assaults from the real world. The will to be or become has been replaced by the will to seem.

8.4 Therapy

At the heart of Adlerian psychotherapy is the process of encouragement, grounded in the feeling of universal co-humanity and the belief in the as yet slumbering potential of the patient or client. By making the patient aware of his secret life plan, the therapist is able to offer an alternative outlook better adapted to the wider world of social interests.

This process of encouragement also makes the Adlerian approach so valuable to all those professions that concern themselves with the development and education of children - therapeutic education being one of Adler's central concerns.

8.5 Continuing Influences

Henri Ellenberger wrote in the seventies of "the slow and continuous penetration of Adlerian insights into contemporary psychological thinking".

Adlerians continue to flourish in the 21st century, some employing an eclectic technique integrating elements of other therapies, from the psychodynamic to the cognitive, others focusing on a more classical approach.

The emphasis on the importance of feelings of inferiority is recognised as isolating an element which plays a key role in personality development. Alfred Adler considered human beings as individuals, hence the reason why he called his psychology "Individual Psychology".

Adler's Individual Psychology and Related Methods

Adler was the first to emphasise the importance of the social element in the re-adjustment process of the individual and who carried psychiatry into the community.

Adler emphasised the importance of equality in preventing various forms of psychopathology, and espoused the development of social interest and democratic family structures for raising children. His most famous concept is the inferiority complex which centres to the problem of self-esteem and its negative effects on human health.

His emphasis on power dynamics is rooted in the philosophy of Nietzsche, whose works were published a few decades before Adler's. However, Adler's conceptualisation of the "Will to Power" focuses on the individual's creative power to change for the better. Adler argued for holism, viewing the individual holistically rather than reductively, the latter being the dominant lens for viewing human psychology.

Adler was also among the first in psychology to argue in favour of feminism, and the female analyst, making the case that power dynamics between men and women (and associations with masculinity and femininity) are crucial to understanding human psychology.

Adler is considered, along with Freud and Jung, to be one of the three founding figures of depth psychology, which emphasizes the unconscious and psychodynamics and thus to be one of the three great psychologist/philosophers of the twentieth century.

9. ADLER'S PERSONAL LIFE

9.1 Earlier Personal life

Alfred Adler was born at Mariahilfer Straße, in Rudolfsheim, a place near Vienna at the time but today part of Rudolfsheim-Fünfhaus, the 15th district of Vienna. He was the second child of seven children of a Hungarian-born, Jewish grain merchant and his wife. Early on, he developed rickets, which kept him from walking until he was four years old.

At the age of four, he developed pneumonia and heard a doctor say to his father, "Your boy is lost". At that point, he decided to be a physician. He was very interested in the subjects of psychology, sociology and philosophy. After studying at University of Vienna, he specialized as an eye doctor, and later in neurology and psychiatry.

Alfred's younger brother passed away in the bed next to him, when Alfred was only three years old. Alfred was an active, popular child and an average student who was also known for his competitive attitude toward his older brother, Sigmund.

In 1895 Adler received a medical degree from the University of Vienna. During his college years, he had become attached to a group of socialist students, among which he had found his wife-to-be, Raissa Timofeyewna Epstein, an intellectual and social activist from Russia studying in Vienna. They married in 1897 and had four children, two of whom became psychiatrists. Their children included writer, psychiatrist and Socialist activist Alexandra Adler, psychiatrist Kurt Adler, writer and activist Valentine Adler and Cornelia "Nelly" Adler.

Author and journalist Margot Adler is Adler's granddaughter.

9.2 Career

Adler began his medical career as an ophthalmologist, but he soon switched to general practice, and established his office in a less affluent part of Vienna across from the Prater, a combination amusement park and circus. His clients included circus people, and it has been suggested that the unusual strengths and weaknesses of the performers led to his insights into "organ inferiorities" and "compensation".

In 1902 Adler received an invitation from Sigmund Freud to join an informal discussion group that included Rudolf Reitler and Wilhelm Stekel. The group, the "Wednesday Society" (*Mittwochsgesellschaft*), met regularly on Wednesday evenings at Freud's home and was the beginning of the psychoanalytic movement, expanding over time to include many more members.

A long-serving member of the group, Adler became president of the Vienna Psychoanalytic Society eight years later (1910). He remained a member of the Society until 1911, when he and a group of his supporters formally disengaged from Freud's circle, the first of the great dissenters from orthodox psychoanalysis (preceding Carl Jung's split in 1914).

This departure suited both Freud and Adler, since they had grown to dislike each other. During his association with Freud, Adler frequently maintained his own ideas which often diverged from Freud's. While Adler is often referred to as "a pupil of Freud's", in fact this was never true; they were colleagues, Freud referring to him in print in 1909 as "My colleague Dr Alfred Adler".

In 1929 Adler showed a reporter with the *New York Herald* a copy of the faded postcard that Freud had sent him in 1902. He wanted to prove that he had never been a disciple of Freud's but rather that Freud had sought him out to share his ideas.

Adler founded the Society for Individual Psychology in 1912 after his break from the psychoanalytic movement. Adler's group initially included some orthodox Nietzschean adherents (who believed that Adler's ideas on power and inferiority were closer to Nietzsche than Freud's). Their enmity aside, Adler retained a lifelong admiration for Freud's ideas on dreams and credited him with creating a scientific approach to their clinical utilization (Fiebert, 1997).

Nevertheless, even regarding dream interpretation, Adler had his own theoretical and clinical approach. The primary differences between Adler and Freud centred on Adler's contention that the social realm (exteriority) is as important to psychology as is the internal realm (interiority).

The dynamics of power and compensation extend beyond sexuality, and gender and politics can be as important as libido. Moreover, Freud did not share Adler's socialist beliefs, the latter's wife being for example an intimate friend of many of the Russian Marxists such as Leon Trotsky.

9.3 Adlerian School of Psychology

Following Adler's break from Freud, he enjoyed considerable success and celebrity in building an independent school of psychotherapy and a unique personality theory. He travelled and lectured for a period of 25 years promoting his socially oriented approach. His intent was to build a movement that would rival, even supplant, others in psychology by arguing for the holistic integrity of psychological well-being with that of social equality.

Adler's efforts were halted by World War I, during which he served as a doctor with the Austrian Army. After the conclusion of the war, his influence increased greatly. In the 1930s, he established a number of child guidance clinics. From 1921 onwards, he was a frequent lecturer in Europe and the United States, becoming a visiting professor at Columbia University in 1927. His clinical treatment methods for adults were aimed at uncovering the hidden purpose of symptoms using the therapeutic functions of insight and meaning.

Adler was concerned with the overcoming of the superiority/inferiority dynamic and was one of the first psychotherapists to discard the analytic couch in favour of two chairs. This allows the clinician and patient to sit together more or less as equals. Clinically, Adler's methods are not limited to treatment after-the-fact but extend to the realm of prevention by pre-empting future problems in the child. Prevention strategies include encouraging and promoting social interest, belonging, and a cultural shift within families and communities that leads to the eradication of pampering and neglect (especially corporal punishment).

Adler's popularity was related to the comparative optimism and comprehensibility of his ideas. He often wrote for the lay public. Adler always retained a pragmatic approach that was task-oriented. These "Life tasks" are occupation/work, society/friendship, and love/sexuality. Their success depends on cooperation. The tasks of life are not to be considered in isolation since, as Adler famously commented, "they all throw cross-lights on one another".

9.4 Emigration

In the early 1930s, after most of Adler's Austrian clinics had been closed due to his Jewish heritage (despite his conversion to Christianity), Adler left Austria for a professorship at the Long Island College of Medicine in the USA. Adler died from a heart attack in 1937 in Aberdeen, Scotland, during a lecture tour, although his cremains went missing and were unaccounted for until 2007.

His death was a temporary blow to the influence of his ideas, although a number of them were subsequently taken up by neo-Freudians. Through the work of Rudolf Dreikurs in the United States and many other adherents worldwide, Adlerian ideas and approaches remain strong and viable more than 77 years after Adler's death.

Around the world there are various organisations promoting Adler's orientation towards mental and social well-being. These include the International Committee of Adlerian Summer Schools and Institutes (ICASSI), the North American Society for Adlerian Psychology (NASAP) and the International Association for Individual Psychology. Teaching institutes and programs exist in Austria, Canada, England, Germany, Greece, Israel, Italy, Japan, Latvia, Switzerland, the United States, Jamaica, Peru, and Wales.

9.5 Basic Principles

Adler was influenced by the mental construct ideas of the philosopher Hans Vaihinger (*The Philosophy of As If / Philosophie des Als Ob*) and the literature of Dostoevsky. While still a member of the Vienna Psychoanalytic Society he developed a theory of organic inferiority and compensation that was the prototype for his later turn to phenomenology and the development of his famous concept, the inferiority complex.

Adler was also influenced by the philosophies of Immanuel Kant, Friedrich Nietzsche, Rudolf Virchow and the statesman Jan Smuts (who coined the term "holism"). Adler's School, known as "Individual Psychology"—an arcane reference to the Latin*individuus* meaning indivisibility, a term intended to emphasize holism—is both a social and community psychology as well as a depth psychology.

Adler was an early advocate in psychology for prevention and emphasized the training of parents, teachers, social workers and so on in democratic approaches that allow a child to exercise their power through reasoned decision making whilst co-operating with others. He was a social idealist, and was known as a socialist in his early years of association with psychoanalysis (1902–1911). *Alfred Adler's Influence on the Three Leading Cofounders of Humanistic Psychology*. Journal of Humanistic Psychology (September 1990)

Adler was pragmatic and believed that lay people could make practical use of the insights of psychology. Adler was also an early supporter of feminism in psychology and the social world, believing that feelings of superiority and inferiority were often gendered and expressed symptomatically in characteristic masculine and feminine styles. These styles could form the basis of psychic compensation and lead to mental health difficulties. Adler also spoke of "safeguarding tendencies" and neurotic behaviour long before Anna Freud wrote about the same phenomena in her book *The Ego and the Mechanisms of Defense*.

From its inception, Adlerian psychology has included both professional and lay adherents. Adler felt that all people could make use of the scientific insights garnered by psychology and he welcomed everyone, from decorated academics to those with no formal education to participate in spreading the principles of Adlerian psychology.

9.6 Approach to Personality

Adler's book, *Über den nervösen Charakter* (*On the Neurotic Character*) defines his earlier key ideas. He argued that human personality could be explained teleologically: parts of the individual's unconscious self ideally work to convert feelings of inferiority to superiority (or rather completeness).[32] The desires of the self ideal were countered by social and ethical demands. If the corrective factors were disregarded and the individual overcompensated, then an inferiority complex would occur, fostering the danger of the individual becoming egocentric, power-hungry and aggressive or worse.

Common therapeutic tools include the use of humour, historical instances, and paradoxical injunctions.

9.7 Psychodynamics

Adler maintained that human psychology is psychodynamic in nature, yet unlike Freud's metapsychology that emphasizes instinctual demands, human psychology is guided by goals and fuelled by a yet unknown creative force. Like Freud's instincts, Adler's fictive goals are largely unconscious. These goals have a "teleological" function. Constructivist Adlerians, influenced by neo-Kantian and Nietzschean ideas, view these "teleological" goals as "fictions" in the sense that Hans Vaihinger spoke of (*fictio*). Usually there is a fictional final goal which can be deciphered alongside of innumerable sub-goals.

The inferiority/superiority dynamic is constantly at work through various forms of compensation and overcompensation. For example, in anorexia nervosa the fictive final goal is to "be perfectly thin" (overcompensation on the basis of a feeling of inferiority). Hence, the fictive final goal can serve a persecutory function that is ever-present in subjectivity (though its trace springs are usually unconscious). The end goal of being "thin" is fictive however since it can never be subjectively achieved.

Teleology serves another vital function for Adlerians. Chilon's "hora telos" ("see the end, consider the consequences") provides for both healthy and maladaptive psychodynamics. Here we also find Adler's emphasis on personal responsibility in mentally healthy subjects who seek their own and the social good.

9.8 Constructivism

The metaphysical thread of Adlerian theory does not problematise the notion of teleology since concepts such as eternity (an ungraspable end where time ceases to exist) match the religious aspects that are held in tandem. In contrast, the constructivist Adlerian threads (either humanist/modernist or post-modern in variant) seek to raise insight of the force of unconscious fictions– which carry all of the inevitability of 'fate'– so long as one does not understand them. Here, 'teleology' itself is fictive yet experienced as quite real.

This aspect of Adler's theory is somewhat analogous to the principles developed in Rational Emotive Behaviour Therapy (REBT) and Cognitive Therapy (CT). Both Albert Ellis and Aaron T. Beck credit Adler as a major precursor to REBT and CT. Ellis in particular was a member of the North American Society for Adlerian Psychology and served as an editorial board member for the Adlerian Journal *Individual Psychology*.

As a psychodynamic system, Adlerians excavate the past of a client/patient in order to alter their future and increase integration into community in the 'here-and-now'.[36] The 'here-and-now' aspects are especially relevant to those Adlerians who emphasize humanism and/or existentialism in their approaches. It also changes the way of how we look at life.

9.9 Spiritual Holism

Metaphysical Adlerians emphasise a spiritual holism in keeping with what Jan Smuts articulated (Smuts coined the term "holism"), that is, the spiritual sense of one-ness that holism usually implies (etymology of holism: from ὅλος holos, a Greek word meaning all, entire, total) Smuts believed that evolution involves a progressive series of lesser wholes integrating into larger ones. Whilst Smuts' text *Holism and Evolution* is thought to be a work of science, it actually attempts to unify evolution with a higher metaphysical principle (holism). The sense of connection and one-ness revered in various religious traditions (among these, Baha'i, Christianity, Judaism, Islam and Buddhism) finds a strong complement in Adler's thought.

The pragmatic and materialist aspects to contextualizing members of communities, the construction of communities and the socio-historical-political forces that shape communities matter a great deal when it comes to understanding an individual's psychological make-up and functioning. This aspect of Adlerian psychology holds a high level of synergy with the field of community psychology, especially given Adler's concern for what he called "the absolute truth and logic of communal life".

However, Adlerian psychology, unlike community psychology, is holistically concerned with both prevention and clinical treatment after-the-fact. Hence, Adler can be considered the "first community psychologist", a discourse that formalized in the decades following Adler's death (King & Shelley, 2008).

Adlerian psychology, Carl Jung's analytical psychology, Gestalt therapy and Karen Horney's psychodynamic approach are holistic schools of psychology. These discourses eschew a reductive approach to understanding human psychology and psychopathology.

9.10 Typology

Adler developed a scheme of so-called personality types, which were however always to be taken as provisional or heuristic since he did not, in essence, believe in personality types, and at different times proposed different and equally tentative systems. The danger with typology is to lose sight of the individual's uniqueness and to gaze reductively, acts that Adler opposed. Nevertheless, he intended to illustrate patterns that could denote a characteristic governed under the overall style of life.

Hence American Adlerians such as Harold Mosak have made use of Adler's typology in this provisional sense:

- The Getting or Leaning.

  They are sensitive people who have developed a shell around themselves which protects them, but they must rely on others to carry them through life's difficulties. They have low energy levels and so become dependent. When overwhelmed, they develop what we typically think of as neurotic symptoms: phobias, obsessions and compulsions, general anxiety, hysteria, amnesias, and so on, depending on individual details of their lifestyle.

- The Avoiding types are those that hate being defeated.

  They may be successful, but have not taken any risks getting there. They are likely to have low social contact in fear of rejection or defeat in any way.

- The Ruling or Dominant type strive for power and are willing to manipulate situations and people, anything to get their way. People of this type are also prone to anti-social behaviour.

- The Socially Useful types are those who are very outgoing and very active. They have a lot of social contact and strive to make changes for the good.

These 'types' are typically formed in childhood and are expressions of the Style of Life.

9.11 Importance of Memories

Adler placed great emphasis upon the interpretation of early memories in working with patients and school children, writing that, "Among all psychic expressions, some of the most revealing are the individual's memories."

Adler viewed memories as expressions of "private logic" and as metaphors for an individual's personal philosophy of life or "lifestyle." He maintained that memories are never incidental or trivial; rather, they are chosen reminders: "(A person's) memories are the reminders she carries about with her of her limitations and of the meanings of events. There are no "chance" memories. Out of the incalculable number of impressions that an individual receives, she chooses to remember only those which she considers, however dimly, to have a bearing on her problems."[41]

9.12 Birth Order

Adler often emphasized one's birth order as having an influence on the style of life and the strengths and weaknesses in one's psychological make up. Birth Order referred to the placement of siblings within the family. Adler believed that the firstborn child would be in a favourable position, enjoying the full attention of the eager new parents until the arrival of a second child. This second child would cause the first born to suffer feelings of dethronement, no longer being the centre of attention.

Adler (1908) believed that in a three-child family, the oldest child would be the most likely to suffer from neuroticism and substance addiction which he reasoned was a compensation for the feelings of excessive responsibility "the weight of the world on one's shoulders" (e.g. having to look after the younger ones) and the melancholic loss of that once supremely pampered position. As a result, he predicted that this child was the most likely to end up in jail or an asylum. Youngest children would tend to be overindulged, leading to poor social empathy.

Consequently, the middle child, who would experience neither dethronement nor overindulgence, was most likely to develop into a successful individual yet also most likely to be a rebel and to feel squeezed-out. Adler himself was the second in a family of six children.

Adler never produced any scientific support for his interpretations on birth order roles, nor did he feel the need to. Yet the value of the hypothesis was to extend the importance of siblings in marking the psychology of the individual beyond Freud's more limited emphasis on the mother and father. Hence, Adlerians spend time therapeutically mapping the influence that siblings (or lack thereof) had on the psychology of their clients.

Adler's Individual Psychology and Related Methods

The idiographic approach entails an excavation of the phenomenology of one's birth order position for likely influence on the subject's Style of Life. In sum, the subjective experiences of sibling positionality and inter-relations are psycho-dynamically important for Adlerian therapists and personality theorists, not the cookbook predictions that may or may not have been objectively true in Adler's time.

For Adler, birth order answered the question, "Why do children, who are raised in the same family, grow up with very different personalities?" While a geneticist would claim the differences are caused by subtle variations in the individuals' genetics, Adler showed through his birth order theory that children do not grow up in the same family, but the oldest child grows up in a family where they have younger siblings, the middle child with older and younger siblings, and the youngest with older siblings. The position in the family constellation, Adler said, is the reason for these differences in personality and not genetics: a point later taken up by Eric Berne.

9.13 Addiction

Adler's insight into birth order, compensation and issues relating the individuals' perception of community also led him to investigate the causes and treatment of substance abuse disorders, particularly alcoholism and morphinism, which already were serious social problems of his time. His wife Dr. Alexandra Adler provided a clinic for the treatment of addicts at the facilities of Professor Pötzl in Vienna.

Adler's work with addicts was significant since most other prominent exponents of psychoanalysis invested relatively little time and thought into this widespread ill of the modern and post-modern age. In addition to applying his individual psychology approach of organ inferiority, for example, to the onset and causes of addictive behaviours, he also tried to find a clear relationship of drug cravings to sexual gratification or their substitutions.

Early pharmaco-therapeutic interventions with non-addictive substances, such as neuphyllin were used, since withdrawal symptoms were explained by a form of "water-poisoning" that made the use of diuretics necessary. Adler and his wife's pragmatic approach and the seemingly high success rates of their treatment were based on their ideas of social functioning and well-being. Clearly, life style choices and situations were emphasized, for example the need for relaxation or the negative effects of early childhood conflicts were examined, which compared to other authoritarian or religious treatment regimens, were clearly modern approaches. Certainly some of his observations, for example that psychopaths were more likely to be drug addicts are not compatible with current methodologies and theories of substance abuse treatment, but the self-centred attributes of the illness and the clear escapism from social responsibilities by pathological addicts put Adler's treatment modalities clearly into a modern contextual reasoning.[44]

9.14 Homosexuality

Adler's ideas regarding non-heterosexual sexuality and various social forms of deviance have long been controversial. Along with prostitution and criminality, Adler had classified 'homosexuals' as falling among the "failures of life". In 1917, he began his writings on homosexuality with a 52-page brochure, and sporadically published more thoughts throughout the rest of his life.

The Dutch psychiatrist Gerard J. M. van den Aardweg underlines how Alfred Adler came to his conclusions for, in 1917, Adler believed that he had established a connection between homosexuality and an inferiority complex towards one's own gender. This point of view differed from Freud's theory that homosexuality is rooted in narcissism or Jung's view of expressions of contra-sexuality vis-à-vis the archetypes of the Anima and Animus.

There is evidence that Adler may have moved towards abandoning the hypothesis. Towards the end of Adler's life, in the mid-1930s, his opinion towards homosexuality began to shift. Elizabeth H. McDowell, a New York state family social worker recalls undertaking supervision with Adler on a young man who was "living in sin" with an older man in New York City. Adler asked her, "Is he happy, would you say?" "Oh yes," McDowell replied. Adler then stated, "Well, why don't we leave him alone." On reflection, McDowell found this comment to contain "profound wisdom", but there must be some misunderstanding on Adler's answer. Adler was offering his help only to those who were asking for it in person. His therapy process could be applied only to those who felt themselves in a deadlock, fallen "at the bottom of a well", and looking for help to get out.

Homosexuality was considered one of the most difficult cases, needing long experience on the part of the psychotherapist and many consequent sessions and much personal work by the individual, depending on the "maturity" of the problem. Success could not be guaranteed.

According to Phyllis Bottome, who wrote Adler's *Biography* (after Adler himself laid upon her that task): "Homosexuality he always treated as lack of courage. These were but ways of obtaining a slight release for a physical need while avoiding a greater obligation.

A transient partner of your own sex is a better known road and requires less courage than a permanent contact with an "unknown" sex. Adler taught that men cannot be judged from within by their "possessions," as he used to call nerves, glands, traumas, drives et cetera, since both judge and prisoner are liable to misconstrue what is invisible and incalculable; but that he can be judged, with no danger from introspection, by how he measures up to the three common life tasks set before every human being between the cradle and the grave. Work or employment, love or marriage, social contact."

9.15 Parent Education

Adler emphasized both treatment and prevention. As a psychodynamic psychology, Adlerians emphasize the foundational importance of childhood in developing personality and any tendency towards various forms of psychopathology. The best way to inoculate against what are now termed "personality disorders" (what Adler had called the "neurotic character"), or a tendency to various neurotic conditions (depression, anxiety, etc.), is to train a child to be and feel an equal part of the family.

The responsibility of the optimal development of the child is not limited to the Mother or Father but to teachers and society more broadly. Adler argued therefore that teachers, nurses, social workers, and so on require training in parent education to complement the work of the family in fostering a democratic character. When a child does not feel equal and is enacted upon (abused through pampering or neglect) they are likely to develop inferiority or superiority complexes and various accompanying compensation strategies.

These strategies exact a social toll by seeding higher divorce rates, the breakdown of the family, criminal tendencies, and subjective suffering in the various guises of psychopathology. Adlerians have long promoted parent education groups, especially those influenced by the famous Austrian/American Adlerian Rudolf Dreikurs (Dreikurs & Soltz, 1964).

9.16 Spirituality, Ecology and Community

In a late work, *Social Interest: A Challenge to Mankind* (1938), Adler turns to the subject of metaphysics, where he integrates Jan Smuts' evolutionary holism with the ideas of teleology and community: "sub specie aeternitatis". Unabashedly, he argues his vision of society: "Social feeling means above all a struggle for a communal form that must be thought of as eternally applicable... when humanity has attained its goal of perfection... an ideal society amongst all mankind, the ultimate fulfilment of evolution."

Adler follows this pronouncement with a defence of metaphysics:

> "I see no reason to be afraid of metaphysics; it has had a great influence on human life and development. We are not blessed with the possession of absolute truth; on that account we are compelled to form theories for ourselves about our future, about the results of our actions, etc. Our idea of social feeling as the final form of humanity - of an imagined state in which all the problems of life are solved and all our relations to the external world rightly adjusted - is a regulative ideal, a goal that gives our direction. This goal of perfection must bear within it the goal of an ideal community, because all that we value in life, all that endures

and continues to endure, is eternally the product of this social feeling."

This social feeling for Adler is Gemeinschaftsgefühl, a community feeling whereby one feels he or she belongs with others and has also developed an ecological connection with nature (plants, animals, the crust of this earth) and the cosmos as a whole, sub specie aeternitatis. Clearly, Adler himself had little problem with adopting a metaphysical and spiritual point of view to support his theories. Yet his overall theoretical yield provides ample room for the dialectical humanist (modernist) and separately the postmodernist to explain the significance of community and ecology through differing lenses (even if Adlerians have not fully considered how deeply divisive and contradictory these three threads of metaphysics, modernism, and post modernism are).

9.17 Death and Cremation

Adler died suddenly in Aberdeen, Scotland, in May 1937, during a three-week visit to the University of Aberdeen. While walking down the street, he was seen to collapse and lie motionless on the sidewalk. As a man ran over to him and loosened his collar, Adler mumbled "Kurt", the name of his son and died. The autopsy performed determined his death was caused by a degeneration of the heart muscle. His body was cremated at Warriston Crematorium in Edinburgh but the ashes were never reclaimed. In 2007, his ashes were rediscovered in a casket at Warriston Crematorium and returned to Vienna for burial in 2011.

10. ADLER'S WORK IN THE 21ST CENTURY

10.1 Absorption into Modern Psychology

Much of Adler's theories have been absorbed into modern psychology without attribution. Psycho-historian Henri F. Ellenberger writes, "It would not be easy to find another author from which so much has been borrowed on all sides without acknowledgement than Alfred Adler."

Ellenberger posits several theories for "the discrepancy between greatness of achievement, massive rejection of person and work, and wide-scale, quiet plagiarism..." These include Adler's "imperfect" style of writing and demeanour, his "capacity to create a new obviousness," and his lack of a large and well organized following.

10.2 Private Logic

A prime modern example of use of Adlerian theory without attribution can be found in the work of American psychologist, Dr. Kevin Leman. Although himself a member of the North American Society of Adlerian Psychology, Leman's self-help book on early memory interpretation only cites Adler for coining the term, "private logic," and neglects to mention that the entire premise for his (Leman's) book—the theory and practice of early memory interpretation—originates with Adlerian Individual Psychology.

10.3 Professional Training

Adler School of Professional Psychology is a post-baccalaureate, non-profit institution of higher education and independent graduate school of psychology located in Chicago, Illinois and Vancouver, British Columbia. As the oldest independent professional school of psychology in North America, the Adler School continues the pioneering work of psychiatrist and first community psychologist Alfred Adler by graduating socially responsible practitioners, engaging communities, and advancing social justice. The Adler School celebrated its 60th anniversary in 2012.

10.4 Adler University

More recently, in November 2013, the Board of Trustees of The Adler School voted unanimously to a name change and to advance the institution's collegiate status to that of a university. The name change -- to Adler University -- will occur officially in January 2015 to reflect both the growth of the school and its broadening pedagogical focus beyond psychology.

The Adler School offers degrees in clinical psychology (Psy.D.) and several master's degree programs, enrolling more than 1,200 students at both campuses. The current president of The Adler School of Professional Psychology is Raymond E. Crossman, Ph.D. He was appointed the fifth president of school in 2003 and since then has realized a new vision, new academic programs, and significant growth.

The Adler School strives to attract applicants to its graduate programs who are broadly interested in social justice -- and its interface with social science, public policy, and the health sciences, rather than applicants who are merely interested in traditional private practice.

10.5 Adlerian Psychology

Adlerian psychology emphasises the human need and ability to create positive social change and impact. Adler held equality, civil rights, mutual respect, and the advancement of democracy as core values. He was one of the first practitioners to provide family and group counselling and to use public education as a way to address community health. He was among the first to write about the social determinants of health and of mental health. Adler's values and concepts drive the mission, work, and values at the Adler School today.

10.6 Community Partnerships

Annually, Adler School students provide over 650,000 hours of community service. The Adler School partners with more than 700 agencies to advance community health. Adler Community

Health Services (ACHS) provides psychological services to underserved populations through its clinical training programs.

ACHS training programs include the Adler Community Mental Health Doctoral Internship in Clinical Psychology as well as psychotherapy and diagnostic assessment externships (practica). The internship is approved by the American Psychological Association (APA) and is a member of the Association of Psychology Post-doctoral and Internship Centers (APPIC).

10.7 Child Guidance Centre

The Adler Child Guidance Centre was established to help care-providers meet the challenges and responsibilities of child guidance. Based on the importance of raising children who are responsible, cooperative, and respectful of self and others, the centre emphasises democratic leadership, encouragement, and reliance upon respectful, non-oppressive, non-coercive methods of teaching discipline. The Centre strives to meet the needs of diverse communities providing cost-effective, non-exclusionary services within the mission and vision of social justice and socially responsible practice.

10.8 Social Exclusion

The Institute on Social Exclusion seeks to analyze the ways in which structural features of society condition human welfare; stimulate public dialogue on the underlying causes of disadvantage and on possible solutions; and engage in practical work that sheds light on and addresses social marginalization.

The institute builds strategic alliances to ensure that all members of society have safe housing, quality education and healthcare, fair terms of employment, nutritious food, personal safety, and judicial equity. We work to dismantle the barriers to these essential rights, opportunities, and resources by advocating for structural change in our society.

10.9 Public Safety and Social Justice

The Institute on Public Safety and Social Justice strives to meet public safety challenges with socially just solutions. They work with community groups, peer institutions, and systems partners to address public safety challenges. By forging creative collaborations, they can devise empirically sound methods beyond mere suppression to create environments where a more lasting and meaningful sense of peace and wellness can prevail.

10.10 Mental Health

The Mental Health and Inclusion Centre advances the wellness and quality of life for sexual orientation and gender variant minorities on an individual and systemic level. They are committed to addressing the needs of those most underserved and underrepresented within these minority groups. Through the application of its principles we aim to improve understanding, inclusion, and service delivery through education, community engagement, and research.

END

INDEX:

	PAGE:
Abreaction and Catharsis	13
Absorption into Modern Psychology	79
Addiction	74
Adler University	80
ADLER'S PERSONAL LIFE	64
ADLER'S PSYCHOLOGY	7
ADLER'S WORK IN THE 21ST CENTURY	79
ADLERIAN APPROACH	61
Adlerian Psychology	80
Adlerian School of Psychology	66
ANALYTICAL CONCEPTS	31
Anxiety	33
Applied Psychology	53
Approach to Personality	69
Aristotle and Works on Psychology	49
Basic Principles	68
Behaviourism	52
Birth Order	73
Breuer, Josef	26
Career Path	5
Career	64
Cathexis	15
Censorship	17
Child Development	44
Child Guidance Centre	81
Child Guidance	9
Child Psychology	45
Clinical Psychology	59
Community Partnerships	80
Comparative Psychology	51
Compensation	61
Constructivism	70
Continuing Influences	62
Cosmopolitan Support	24
Death and Cremation	78
Depth Psychology	14
Developmental Psychology	56
Disagreement	39
Earlier Personal life	64
Early Life and Career	26
Educational Psychology	57
Ego	36
Experimental Psychology	55
FORERUNNERS AND CONTRIBUTORS	26

83

Freud's Sexuality and Development	20
Goal and Development	6
History of psychology	48
Homosexuality	75
Hysteria Disorder	32
Hysteria	12
Id	35
Importance of Memories	72
INDIVIDUAL PSYCHOLOGY	10
Industrial Psychology	55
Inferiority Complex	11
Inferiority Complex	5
Inferiority	7
Initial Differences	7
Interest in Psychology	50
Jung, Carl	26
Jung's Character of Psychotherapy	27
Levels of Consciousness	40
Libido	35
Logic, Philosophy of Psychology	48
Mental Health	82
Mental Hygiene	43
Mental Topography	16
Neo-Freudianism	6
Normality	10
Oedipus Complex	18
Oedipus	37
PARALLEL CONTRIBUTIONS	12
Parapraxes	16
Parent Education	76
Personality Development	8
Phallic Stage	21
Physician and Patient	14
PIONEER MEDICAL SPECIALIST	5
Pleasure and Pain Principle	15
Practice and Theory	5
Private Logic	79
Professional Training	79
Psychiatry	42
Psychoanalysis	12
Psychoanalysis	31
Psychoanalytic Dreams	38
Psychoanalytic Movement	19
Psychodynamics	69
PSYCHOLOGY AND BRANCHES	47
Psychopathology	58

Psychotherapies	41
Public Safety and Social Justice	81
RANK, OTTO	29
Scientific Discipline	47
Self-realisation	8
Sense and Sensible	49
Sexual Development	22
Sexual Instincts	17
Sigmund Freud	12
Social Exclusion	81
Social Psychology	56
Social Psychology	58
Spiritual Holism	70
Spirituality, Ecology and Community	77
Subject Matter	13
Sublimation	19
Superego	36
Theoretical Basis	16
Theories	10
Therapy	62
Transference	18
Typology	71
Unconscious	40
Withdrawal	61

Adler's Individual Psychology and Related Methods

BIBLIOGRAPHY

All books publications mentioned in this bibliography are written by Andreas Sofroniou

1. MEDICAL ETHICS THROUGH THE AGES, ISBN: 978-1-4092- 7468-1
2. MEDICAL ETHICS, FROM HIPPOCRATES TO THE 21ST CENTURY ISBN: 978-1-4457-1203-1
3. THE MISINTERPRETATION OF SIGMUND FREUD, ISBN: 978-1-4467-1659-5
4. JUNG'S PSYCHOTHERAPY: THE PSYCHOLOGICAL & MYTHOLOGICAL METHODS, ISBN: 978-1-4477-4740-6
5. FREUDIAN ANALYSIS & JUNGIAN SYNTHESIS, ISBN: 978-1-4477-5996-6
6. PSYCHOTHERAPY, CONCEPTS OF TREATMENT, ISBN: 978-1-291-50178-0
7. PSYCHOLOGY, CONCEPTS OF BEHAVIOUR, ISBN: 978-1-291-47573-9
8. PHILOSOPHY FOR HUMAN BEHAVIOUR, ISBN: 978-1-291-12707-2
9. SEX, AN EXPLORATION OF SEXUALITY, EROS AND LOVE, ISBN: 978-1-291-56931-5
10. PSYCHOLOGY FROM CONCEPTION TO SENILITY, ISBN: 978-1-4092-7218-2
11. PSYCHOLOGY OF CHILD CULTURE, ISBN: 978-1-4092-7619-7
12. JOYFUL PARENTING, ISBN: 0 9527956 1 2
13. THE GUIDE TO A JOYFUL PARENTING, ISBN: 0 952 7956 1 2
14. THERAPEUTIC PHILOSOPHY FOR THE INDIVIDUAL AND THE STATE, ISBN: 978-1-4092-7586-2
15. PHILOSOPHIC COUNSELLING FOR PEOPLE AND THEIR GOVERNMENTS, ISBN: 978-1-4092-7400-1
16. MORAL PHILOSOPHY, FROM HIPPOCRATES TO THE 21ST AEON, ISBN: 978-1-84753-463-7
17. MORAL PHILOSOPHY, THE ETHICAL APPROACH THROUGH THE AGES, ISBN: 978-1-4092-7703-3
18. ARISTOTLE'S AETIOLOGY, ISBN: 978-1-4716-7861-5
19. SOCIOLOGY, CONCEPTS OF GROUP BEHAVIOUR, ISBN: 978-1-291-51888-7
20. SOCIAL SCIENCES, CONCEPTS OF BRANCHES AND RELATIONSHIPS ISBN: 978-1-291-52321-8
21. CONCEPTS OF SOCIAL SCIENTISTS AND GREAT THINKERS, ISBN: 978-1-291-53786-4

www.ingramcontent.com/pod-product-compliance
Lightning Source LLC
Chambersburg PA
CBHW072232170526
45158CB00002BA/857